S0-BUG-228

Managing Millions

Managing Millions

An Inside Look at High-Tech Government Spending

Rodney D. Stewart
and
Ann L. Stewart

WILEY

A Wiley-Interscience Publication

JOHN WILEY & SONS

New York Chichester Brisbane Toronto Singapore

Library of Congress Cataloging in Publication Data:

Stewart, Rodney D.
Managing millions: an inside look at high-tech government spending/Rodney D. Stewart, Ann L. Stewart.
p. cm.
"A Wiley-Interscience publication."
Bibliography: p.
Includes index.
ISBN 0-471-84709-7
1. Government purchasing—United States. 2. Government purchasing—United States—Cost control. I. Stewart, Ann L. II. Title.
JK1673.S74 1988
353.0071'2—dc19 88-15513
CIP

Printed in the United States of America

10 9 8 7 6 5 4 3 2 1

In memory of
John Russell "Russ" Clark
1908–1986

Contents

Preface

Some of the largest chunks of the U.S. federal budget are consumed by major systems and projects. One of the most perplexing problems of our time is how to improve the three *e*'s (efficiency, economy, and effectiveness) in these major programs. Panels, commissions, and committees have investigated these big systems and how they were procured; expert consultants have been hired to give advice on both technical and management aspects of programs; and research corporations and universities have been used to suggest new procurement, development, and production strategies. Yet fundamental problems still exist in being able to (1) predict costs of major systems and projects; (2) produce reasonably priced, effective systems and projects; and (3) carry out development programs and systems that meet original expectations. System developers have often failed to heed the advice of technical and management people in their own organizations as well as outside consultants: Political or profit factors have often taken precedence over sound engineering and management principles in acquiring programs.

Few people realize that the same system of government procurement that resulted in the procurement of $500 hammers is being used to buy $1 billion submarines. Overspecification and inadequate cost control cause big programs as well as spare parts to be overpriced, and the potential savings and cost avoidance in big programs are enormous. Spare parts overpricing represents only the tip of the iceberg when it comes to potential savings and opportunities for economical and effective procurement in major systems. The problems and the potential fixes are well known to those involved in the programs, but strong counterforces that resist change and foster inefficiency and waste continue to maintain a persistent foothold among the myriad organizations and companies that supply the nation's major programs.

Despite much experience in procuring systems of all sizes and types, there are still major problems in how we acquire many of our big programs. These problems are manifested principally in (1) continued failure to meet schedule and performance goals and (2) constant escalation of costs above original estimates. As a result of these two problems, fewer systems and projects are initiated and carried out than would be possible with more accurately estimated and more efficiently managed programs. We know good programs are possible (in terms of setting and meeting cost, schedule, and performance goals) because examples of successful programs do exist. This book analyzes a few of these successful programs in layman's language, reviews top studies that have recommended big changes that could produce beneficial results, and delves into some of the underlying principles as to why more effective action has not been taken in making long overdue improvements. This book synthesizes the results of these studies and suggests a framework of actions required to reform our acquisition process. Finally, the book encourages all citizens to become involved and informed as to the benefits, advantages, and actions to be taken to achieve positive results and to write, call, or meet with their representatives or senators to encourage speedy reform.

RODNEY D. STEWART
ANN L. STEWART

Huntsville, Alabama
August 1988

Acknowledgments

We would like to express our appreciation to all of those who graciously provided their time, advice, and suggestions in helping us research this book. We particularly appreciate the time spent by those we interviewed and for their permission to use excerpts from the interviews. Among those who have given permission to quote them directly in the manuscript are Dr. Eberhard Rees, former director of NASA's George C. Marshall Space Flight Center; Mr. Larry Seggel, project manager for the Army's Multiple Launch Rocket System; Mr. David Dukes, former director of the Joint Financial Management Improvement Group; Mr. James Gregory, former deputy chief of the Contracts Office of the Army's Strategic Defense Command; and Mr. Fletcher Lutz, executive director of the Association of Government Accountants. Others who contributed valuable information and insight into the government's system and project acquisition process were Mr. Mike Motley, Mr. Tim Desmond, and Mr. Les Farrington of the General Accounting Office, Mr. Dave Baker of the Office of Federal Procurement Policy, Messrs. Al Berman, Dave Gribble, and Bill McQuaid of the Office of Management and Budget; Mr. Dick Reynolds of the Defense Advanced Research Projects Agency; Ms. Colleen Preston and Mr. Rudy DeLione of the House Armed Services Committee; Mr. John Etherton of the Senate Armed Services Committee, Senator Howell Heflin and Congressman Ronnie Flippo. We wish to thank Mr. John McAlheny, Mr. Dave Shilling, Mr. Steve Bhattacharya, and Mr. Al Long of the Metropolitan Washington Mass Transit Agency for providing technical, financial, and on-site information about the Washington, D.C. metrorail project. We are also indebted to Mr. Burt Hall, Mr. Don Sowle, Dr. Bill Wall, Mr. Ken Hartley, and Retired General George Sylvester for pointing out pitfalls and recommending changes in early drafts of the chapters on national planning, procurement, and project management. We are especially grateful for the assistance provided to us by Mr. Mike Adcock, staff assistant to Congressman Ronnie Flippo, who kept information flowing to us from sources in Congress, the Congressional Research Service, the Library of Congress, and the committee staffs in both the Senate and the House of Representatives. We also appreciate the help given to us by Mr. Larry Margolis, professor of political science at the University of Alabama in Huntsville; Mr. Harry D. Cleaver, a consultant in government contract management; Mr. Joe Walton of the Naval Special Projects Office for an interview; Mr. Ed Yates of Applied Research, Incorporated, for helpful comments; and the National Contract Management Association for permission to reprint quotes from *Contract Management* magazine.

Finally, we appreciate the friendly support and continued encouragement provided by Mr. Frank Cerra, editor at John Wiley & Sons, the excellent comments by Wiley's reviewers, and the highly professional book marketing and production staff at Wiley.

Managing Millions

CHAPTER ONE

Major Systems in the Age of High Technology

Two hundred feet below a shopping center in Wheaton, Maryland, an Austrian engineer is directing his crew in the placement of a special absorbent layer and vinyl shield on the inside of an enormous tunnel hollowed out of rock near the U.S. capitol. His work is part of a $9.4 billion project to build a 109-mile high-tech rapid transit system to carry commuters to and from suburban areas near Washington, D.C. Eighty percent of the project is federally funded.

Somewhere in the far north, on the fringes of the Arctic Circle, an Air Force lieutenant is peering at the display of an advanced, high-tech early warning system. Her job is to operate a multimillion-dollar, government-purchased advanced radar system that detects hostile space satellites, incoming aircraft, or enemy missiles.

Far to the south, at the Kennedy Space Center in Florida, technicians are meticulously checking the joints on the immense rockets developed by NASA that will boost the next and crucial space shuttle vehicle into orbit around the earth.

And far out in the Pacific, on a submarine deep under the ocean's surface, a seaman has just finished the last step of an exacting software checkout procedure that will permit an underwater launch test of the Navy's newest Cruise missile.

These people, and tens of thousands of others throughout the United States—and throughout the world—are engaged in some of the largest and most expensive tasks ever undertaken: the development, testing, and operation of major systems funded wholly or partially by the federal government.

What is a major system? Why is a major system different from other human undertakings, and why are major systems receiving so much attention from the press, the Congress, and the U.S. public?

In this century, major systems are among the principal tools we have chosen to help accomplish the objectives our forefathers set out in the U.S.

Constitution: to form a more perfect union, to establish justice, to ensure domestic tranquility, to provide for the common defense, to promote the general welfare, and to secure the blessings of liberty for ourselves and our posterity.[1] These systems take the form of civil construction projects, defense systems, space systems, weapons systems, and information systems. As we move from the industrial age into the high-tech information age, these systems have become more complex, more costly, and more time-consuming to develop, test, build, and operate.

A system is "an assemblage or combination of elements or parts forming a complex or unitary whole."[2] There are natural systems such as river systems and ecology systems, and there are man-made systems such as transportation systems, weapons systems, communications systems, and information systems. The colonists who came to our shores in the 1600s discovered a great wealth of natural resources, but they soon found a need to build civil systems such as roads, bridges, and canals. One of the first major civil systems built in this country was a canal between the Charles and Neponset rivers in Dedham, Massachusetts. It was completed prior to 1640.[3] Weapons on land consisted almost exclusively of the flintlock musket with bayonet, but on the high seas, the colonies found that protection was needed against pirates. And in 1631 the English colonists in Massachusetts built what was probably the first American major system, the thirty-ton warship *Blessing of the Bay*, which marked the birth of the Colonial Navy. In 1775, two merchant ships were converted into combat vessels. These represented the first major fighting systems of the Continental Navy. By the late 1700s the United States had built hundreds of ships including privateers, or privately owned vessels. Expanding business, commerce, and trade required many new civil construction projects.

Today, major man-made systems have become multimillion- and multibillion-dollar projects which take years to develop, involve thousands of elements or parts, and affect virtually every phase of our life and economy. It is these complex, multiyear, megadollar projects that we define as major systems. The accepted definition of a major system in the Department of Defense, according to DOD Directive 5000.1, includes cost boundaries of more than $200 million for development and $1 billion for procurement; but any system that requires massive monetary, human, and technical resources can be classified as "major."

Because major systems are complex and require such extensive resources, there are myriad ways in which enormous savings can be achieved through efficient acquisition, just as there are a multitude of ways in which money and human effort can be squandered if effective and economical methods are not used. Major systems are the most visible of all products of our government because they provide tremendous benefits; but if not well conceived and managed, they voraciously and promiscuously devour resources.

The policies, practices, and procedures used to conceive and procure major civil, defense, and space systems are similar to those used to manage other activities within and outside of government. The successes and failures in major acquisitions point the way toward achieving the three *e*'s (efficiency, economy, and effectiveness) in other areas of our government and in industry as well. The value of a searching analysis into major systems acquisition, therefore, has far-reaching effects, potential long-term benefits, and penetration into phases of procurement beyond just the highly visible megaprojects.

Very few systems in George Washington's time took more than a year or two to build. The sail-driven warships of the time, which took two to four years to build, were notable exceptions. There were no huge research and development teams generating thousands of ideas, inventions, patents, and scientific breakthroughs as there are now. The military used existing technology for the large part, adapting and modifying what was already available to supply the Continental Armies. At the time there was little need for multiyear projects or multiyear planning. (The cannon lock, a major advancement in artillery weapons invented and patented by Lt. Dahlgren in 1835, however, took 15 years to be incorporated into an operational weapon.[4]) The country responded to needs as they arose. And little was known about systems engineering, project management, or high-technology methods.

Foundations for the major systems and projects of today were laid as early as the 1700s in the emerging industrial revolution, when domestic production evolved into industrial production. Industrial production required enormous outlays of money for land, buildings, machinery, and power. The early factories, power plants, and dams were major systems of the era. As technology advanced into the 1800s and 1900s, the size and complexity of machinery and factories increased. The aerospace industry got started in the 1900s with the birth of powered flight, and the Manhattan Project brought the nuclear age into being. In the postwar era, electronics came of age, paving the way to the information and computer age of the 1980s and 1990s.

As civil construction projects and military and space systems emerged and became more complex, a number of engineering disciplines and "ilities" (producib*ility*, reliab*ility*, affordab*ility*, supportab*ility*, and such) emerged that were necessary for the development and use of the products, projects, and services needed to provide for the common defense and promote the general welfare. These engineering disciplines, and the increasing technical content of the projects themselves, required highly trained managers, engineers, scientists, and technicians. Transforming ideas into operational systems took years, and the expense of the development and investment process began to grow.

As the complexity and technical content of major systems increased, so did the problems: problems in keeping jobs on schedule and within acceptable cost boundaries and in maintaining technical and user requirements. Horror stories began to emerge as systems' budgets grew. A profusion of major systems procurement and management problems has continued despite large expenditures of money and time on studies; legislation; management systems; personnel; reorganizations; and revised policies, procedures, and practices.

Yet there have been many *successful* projects that have been completed on time and within original cost estimates. What made these successful projects different? Why did some projects succeed while others failed? If some projects were successful, why couldn't *all* projects be successful? These questions have been asked in a number of government-funded studies over the past several decades. In these studies the underlying difficulties in major systems acquisitions have been identified over and over again. Some of the recommendations of these studies have been heeded—others have not. Yet the same problems have recurred. What is wrong? Why haven't these studies, commissions, blue-ribbon panels, investigations, and audits borne fruit in terms of producing successful project outcomes? Why haven't significant beneficial changes been made in the systems acquisition process? Why have we been unable to successfully and durably implement changes along these lines? Has it simply never occurred to anyone to actually carry out these recommendations or are there powerful forces or human factors that are working against beneficial reform?

There have been dozens of major studies and reports prepared in the past four decades[5] recommending changes to improve efficiency, economy, and effectiveness in the Department of Defense alone. Of these, over half pertain to the major systems acquisition process. These studies were performed by specially appointed committees (1953, 1962), blue-ribbon panels (1970, 1986), professional societies and research institutes (1969, 1970, 1985), the Congressional Research Service (1975, 1984), private sector groups (1983), not-for-profit corporations under contract (1986), the General Accounting Office (1976, 1983, 1984, 1986), and the Department of Defense (DOD) (four major studies in the mid-1970s, followed by studies in 1980, 1981, 1983, and 1984). Of the DOD studies in the mid-1970s, three were performed by the military services and one by the Office of the Secretary of Defense. A special one on naval ship-building was performed because of contractor claims of an overrun of $1 billion. Were these studies simply intellectual exercises for generations of policy analysts, or can they be effectively used to forge major policy changes that will guarantee the economy and effectiveness of our major systems?

The reports and studies of Department of Defense major systems, as well as broader studies of government procurement in general, have resulted in myriad changes in organizations, policies, procedures, and laws regulating and controlling the major systems acquisition process. Some changes, as we will show in later chapters, have had detrimental rather than beneficial effects on the procurement process. Certain fundamental changes that have been needed since the inception of cost-reimbursement contracting have been avoided; and some recommended changes that could result in solving deep-seated problems and effecting far-reaching improvements have been ignored. In the meantime, big problems in procuring big systems continue.

The Commission on Government Procurement

The largest single study of government procurement undertaken in recent years was done by the Congressionally established Commission on Government Procurement in 1969 through 1972.[6] The study, enacted through legislation by Congress (Public Law 91-129) after a series of widely publicized cost increases and program slippages in major defense and civil acquisitions in the 1960s, covered all aspects of government procurement.

The $7 million study (not counting the salaries of the 500 persons loaned to the Commission by high-tech industries and government agencies) resulted in 15,000 pages of reports that embodied some 149 recommendations. The recommendations advocated specific legislation and policies for implementation by Congress and the executive branch. Many of the 149 recommendations of the Commission were implemented, yet problems in procurement and in major systems acquisition continue. Were they the right recommendations? Were the resulting policies helpful or were some of them actually *counterproductive* to the stated goals of improving efficiency, economy, and effectiveness in government procurement? What were the findings of the study, and were these findings accurately translated into recommendations and subsequently fully implemented?

When the term "procurement" is used in government, it is often related to the purchasing or contracting function rather than the overall management function. Hence, the overriding emphasis of many of the Procurement Commission's study groups was on methods of contracting, legal remedies, procurement legislation, procurement regulations, and procurement organizations. But the Procurement Commission did not limit its reform suggestions to the procurement and contracting fields. For example, in major systems acquisition, it made broad recommendations ranging from budgetary reforms and personnel issues to agency organizations and the Congress itself. The Procurement Commission enlisted the help of thirteen study groups and several special teams. These groups and teams, which were the source of in-depth findings, addressed some 450 problems and issues in government procurement that had been consolidated into 325 study topics.

Three of the study groups on the Procurement Commission were to concentrate on *types* of procurement: research and development, major systems, and commercial products. The Major Systems Acquisition Study Group, headed by Russ Clark, then executive vice president of LTV,[7] had a broad charter that included a study of planning, budgeting, and funding methods; national policy issues affecting the major systems acquisition process; and systems requirements definition and initial acquisition planning.

The Major Systems Acquisition Study Group[8] consisted of eleven representatives of government agencies including NASA, the Office of Management and Budget, the General Accounting Office, the U.S. Air Force, and civil agencies such as the Department of Transportation and the Department of Health, Education, and Welfare. The list of nongovernment organizations represented (also eleven members) read like a "who's who" of aerospace and defense contractors. Aerojet General, the Aerospace Corporation, Boeing, General Electric, Lockheed, North American, and Western Electric as well as LTV Aerospace were represented. Also included in the nongovernment side were representatives of two industry groups: the Aerospace Industries Association and the Shipbuilder's Council of America.

In one of the most intensive travel and interview schedules ever undertaken, members of the major systems study group interviewed 572 persons over the four-month period of April through July 1971. Persons interviewed included those in government, industry, the academic community, and foreign governments. The key question posed by study group members was, "What is the reason for the current large overruns and cost increases in major civil and military projects, and how can these cost increases be eliminated in future projects?" The responses were surprisingly consistent and formed a pattern for a set of study group findings. Prior to and since 1971, the same pattern of findings has emerged from other similar study groups—from the Rockefeller Committee in 1953 to the Packard Commission in 1986.

Principal findings of the study group, spelled out in its initial draft[9] and later condensed into the final procurement commission report,[6] listed among the causes of inefficiency in major systems procurement: premature program commitment, optimism of system advocates, uncertain and delayed funding, and unstable program requirements. The commission's report and backup material contain excellent material and supporting rationale for three of the key recommendations of this book: the use of phased or incremental development, natural rather than forced competition, and Congressional commitment to an orderly major systems planning and approval process. Findings in the major systems realm included overly complicated source selection procedures, restricted project manager authority, and excessive management layering. It is interesting to note that most of the Procurement Commission's findings in the area of major systems acquisition were concerned with the management and funding process rather than the procurement process. Herein lies the one reason that the Procurement Commission recommendations did not result in reforms that would correct the then growing and still prevalent underlying ailments in the way the government performs its major projects. Maladies in the major systems acquisition process cannot be corrected by contract language, procurement incentives, legislation, or reorganizations. The medicine must be applied at the source of the disease, which is *in the funding and management of these projects*. And the responsibility for the cure goes to the highest levels of government, even to the President himself—and then to the Congress, which must represent the taxpayers who are the ultimate customers and beneficiaries of major systems.

Findings in the major systems area were not "procurement-oriented"; that is, they could not easily be expressed in procurement legislation, procurement regulations, or contract language. Further, the Procurement Commission's results were given much less publicity than the problems that created its formation. In the law that extended the Commission's activities through December 1972, the House and Senate Government Operations Committees stated that the Commission should "furnish the Congress and the President with a blueprint for future federal procurement policies and practices." The blueprint delivered was 149 recommendations provided to the Congress. Less than half of these recommendations were carried to full original intent. While only a small number were rejected outright, many that were adopted received no substantive action and others remained in limbo. A number of inconsistencies in the procurement regulations and policies of the federal government were corrected, but the deep-seated, long-standing problems of program cost increases, schedule slippages, and performance shortfalls continued.

Reorganization Itself is Not a Cure

A commonly used generic cure for acquisition management inadequacies is *reorganization*. Roman satirist Gaius Petronius wrote in the first century A.D. that he was to learn later in life that one tends to meet any new situation by reorganizing; he said that it can be a wonderful method for creating an illusion of progress while producing confusion, inefficiency, and demoralization. Reorganization is not bad in and of itself; in fact it is good practice to permit free and regular reorganizations to occur when innovation and creativity are key elements. Tom Peters and Bob Waterman point this out in their book, *In Search of Excellence*.[10] They advocate three "pillars" in a management structure: one that represents stability, one that represents entrepreneurship, and one that encourages the breaking of old habits.

Reorganization through legislation, however, is frequently involved in attempts to achieve reform, improvement, or change in the federal government. Reorganization, however, provides no benefit unless it is accompanied by fundamental changes in management philosophies. A reorganization without redirection of goals merely redistributes the same problems to different people. Even major reorganizations, such as that enacted in the Defense Reorganization Act, garnered only half-hearted support by military observers and by Congress. One Washington correspondent said that the reorganization should force the branches to more closely agree on what really is needed to meet the threat and, hopefully, avoid some duplication that has taken place. This type of thinking, although it proceeds in the right direction, does not go far enough to result in fundamental changes or long-term improvements. One senator said he voted for the bill because passage was inevitable and he didn't want to hurt Senator [Barry] Goldwater's feelings. This apparent apathy and indifference to the real chances for success in such a nationally important measure are symptomatic of the atmosphere of "Let's try something else—a new organization—and maybe it will work better than what we have had before." In a study of the advantages and disadvantages of a centralized civilian acquisition agency,[11] the General Accounting Office found that "acquisition problems were not necessarily organizationally related and no single action, such as establishing a centralized civilian agency, would solve them."

The penchant for using a reorganization as a cure-all for the major systems acquisition process has been commonplace for many years in industry as well as government. For example, an interview with James Gregory, former Deputy Chief of the Contracts Office for the Army's Safeguard Missile Systems Command (now the Strategic Defense Command) and Chief of the Contract Cost Division of the Army Missile Command,[12] included the following interchange. (The interview starts just after Gregory has told us about several successful projects—a subject that will be discussed further in Chapter 2.)

ROD: *Can you single out a program that wasn't quite as successful?*

ANNIE: *The other end of the spectrum?*

JIM: *Of course some programs were just terminated.*

ROD: *Like what?*

JIM: *I'm trying to think of what's the most recent one we had . . . Viper*

ROD: *Viper? Was it terminated early in the program or. . .*

JIM: *No, it was in production.*

ROD: *So there was a lot of money already spent?*

JIM: *The development was spent and we had production contracts. It couldn't meet cost or performance objectives.*

ROD: *What do you think that was caused by?*

JIM: *You know there are two sides to the fence. The contractor contended that the development objectives and production objectives were just not obtainable. The government, of course, contended that it could have been done if we'd had a better managed program. And there is evidence that the contractor had a number of management teams work on it.*

ANNIE: *It must have had a number of project managers then?*

JIM: *Oh, yes. That's often the contractor's answer. That's one problem I have with contractors. Their answer to everything is we'll just get a new management team. When something happens or goes bad, they get rid of those managers and get some more and that's very expensive.*

ROD: *Instead of resolving the problem?*

JIM: *Yes, instead of resolving the problem, they take the baseball attitude, fire the manager and get new managerial teams, new players or something.*

ROD: *Or a new coach for the football team?*

JIM: *Yes, right. That doesn't always work, but that's the contractor's remedy very often.*

ROD: *And that involves dollars too, doesn't it? Delay and reeducation of a new team?*

JIM: *Yes. The people doing the work don't quite know what to do for a while because they have a new management team.*

Jim's experience parallels our own in project management, that a new manager or a new team is often *not* the answer to solving a major procurement or system problem (more about this when we talk about project managers and life-cycle management).

But clearly, reorganization alone is not a cure. And government reorganization often results in more layers of management because an old organization dies hard and the new organization is often not given the complete authority to do the job.

"Sparks" Hiestand, a highly respected, long-term defense contracting official, now a counsel to the law firm of Morgan, Lewis and Bockius and a member of the Procurement Round Table Board of Directors, wrote in a recent issue of *Government Executive*[13]: "Lurking in the background was the Office of Federal Procurement Policy (OFPP), created by Congress in 1974, which was supposed to exercise overall control over agency procurement policies and procedures. [The formation of the Office of Federal Procurement Policy was one of the 149 recommendations of the Commission on Government Procurement.] Its attempts to do so were frustrated by subsequent limitations imposed by Congress." Although the Office of Federal Procurement Policy turned out many useful functions, this was another example of an unsuccessful attempt to solve problems by setting up a new organization.

The Office of Under Secretary of Defense for Acquisition was recently established to exercise what *Contract Management* magazine (March 1987) termed "Broad powers over the DOD procurement process," but without control of operational test and evaluation, a program element that GAO and others have found to be essential for program success. There have even been groups that have been advocating a separate testing organization not under the control of the project manager (this would take away any remaining ability of the project manager to develop a successful system) or the establishment of a completely autonomous civilian acquisition agency to procure all U.S. major systems. But history has shown that setting up new offices that do not have the proper authority merely results in the layering of management controls over the procurement process rather than a streamlining of organizations and procedures.

Examples of layering of management in the procurement process include establishment of competition advocates in every service as a result of the Competition in Contracting Act and the establishment of acquisition executives in each branch of the armed services as recommended by the 1986 Packard Commission report. These advocates and executives, each with their own staff and set of regulations, were placed in addition to, rather than in place of, existing functions. Our appeal is for the government to make fundamental improvements in the major systems acquisition process *but to do it with the people and organizations that already exist* rather than by establishing new offices, agencies, advocates, executives, committees, or commissions. We advocate simply pointing existing people and organizations in a new direction rather than establishing whole new sets of personnel, procedures, regulations, and controls. For example, the Office of Federal Procurement Policy could become a much more effective agent for procurement reform if it had a say-so in personnel assignment, management, funding, and other matters as well as in procurement matters.

One might reasonably ask the question: "Since the Packard Commission (and other government-wide and DOD study groups) tried to reduce the layers of management in the acquisition process, exactly why and how did it fail?" It failed principally because it was trying to accomplish its objectives *by adding onto existing organizations* rather than by making the deep-seated, fundamental changes needed to assure lasting success. Changes were recommended as if there were no limitations on resources, personnel, or organizational size; and there was not a mandate at the outset to recommend only improvements that could be accomplished within existing resources.

At this writing, another new office, the Office of Profit Studies and Review (OPSR) is being proposed by GAO and others to study and monitor contractor profit. The GAO report that proposes the new office does not indicate which function in the federal government it would replace or why the function cannot be performed by an existing organization. Other new organizational elements such as a Value Engineering Council, ethics committees, and profit-monitoring teams are proposed on a regular basis. In summary, legislated reorganization without accompanying changes in fundamental management philosophy is ineffective, inflexible, and expensive. It creates more bureaucracy without solving the deep-seated and pervasive problems it was created to address. And once an organization is created by federal law, it is very difficult to dissolve.

What the Experts Say

The experts (and many nonexperts) have lots to say about major systems. The 1971–1972 Procurement Commission's major systems study group collected 310 documents, which included government reports, books, news articles, speeches, Congressional hearing transcripts, industry briefs, research papers, theses, and presentations—all related to the acquisition of major systems. Since that time, we have obtained 200 more documents to provide the latest information on major systems policies and practices (see Bibliography). Reams of written material on the acquisition of major systems exist in government, private industry, and university libraries. Many more such documents are being created each day by Congressional staffs, the General Accounting Office, agency auditors, research institutes, universities, and by private industry under government contract. The key to solving the major systems puzzle is to search for the common thread in all of these documents and expert opinions. What are the things that are being said uniformly and consistently? Plowing through the smokescreen of controversy and academic mumbo-jumbo, what are the key issues and the principal or fundamental changes or improvements that are needed? And, in searching for the common ground and the areas where all agree, can we also find some information on *how* to make constructive changes in the megasystems acquisition process?

To address some of these questions, and to further update our data base for this book, we interviewed officials in the Office of Management and Budget, General Accounting Office, Army, Navy, and Air Force, and NASA to see if the 149 recommendations made in 1971 by the Procurement Commission; the various laws, regulations, and policies put forth since then; and the Packard Commission recommendations have had any effect on the three *e*'s (efficiency, economy, and effectiveness) of major systems acquisition. The interviews showed that many, if not most, of the same fundamental problems investigated by the Procurement Commission in the early 1970s still exist.

According to Hiestand,

> *. . . the Commission on Government Procurement verified there were serious problems and made recommendations to solve them, including simplification and clarification of the existing procurement laws. Although a number of bills were introduced in Congress in the 1970s to consolidate and restate the procurement statutes to express policy rather than procedures, they withered and died, partly because of opposition by the Department of Defense. Time passed and the statutory framework became more fragmented, the bureaucracies grew, and the implementing regulations became more diverse and uncoordinated.*[13]

Some officials are pessimistic that any complete solution to acquisition problems exists; yet others have told of remarkably successful programs that have been completed on schedule and within original cost estimates. Experts or authorities on major systems acquisition in government—joined by those in the private, academic, and public sectors who have studied the major systems acquisition process, government procurement, and project management in depth—have developed fundamental principles required for success. Among these are the many commissions, panels, and study groups previously mentioned; professional groups such as the Project Management Institute (Drexel Hill, Pennsylvania) and the Large Scale Programs Institute (Austin, Texas); and authoritative authors and educators on the subject. These experts say that it *is* possible to have a well-conceived, -established, and -managed project and that certain principles, if conscientiously applied, will invariably result in successful programs.

Industry, too, has shown that it recognizes and can achieve success. The 1985–1986 rash of "excellence" books on methods of developing excellence in a company were virtually all best sellers because people have faith in principles and approaches that have been successful. The theory is that excellence can be achieved by synthesizing and applying the inputs from the experts based on areas of agreement and lessons learned. There is now strong evidence that this theory can be fruitfully put into practice. The things we learned in our recent interviews were remarkably consistent with what we can synthesize from these experts and what we heard 16 years earlier in the Major Systems Study Group's interviews. The largest complaints revolve around the inability to tie down project funding long enough to get a major project phase completed; the absence of a long-term, integrated plan for major systems; instability in project managers' assignments; excessive layering of additional requirements on projects; concurrent rather than evolutionary, phased development; too many procurement laws; and absence of a good accounting system in government.

On April 5, 1976, the director of the Office of Management and Budget issued Circular #A109: Major Systems Acquisition. This circular, formulated by the Office of Federal Procurement Policy (OFPP), was designed to establish policies to be followed by *all* executive branch agencies in the acquisition of major systems. The policy was intended to effect reforms that would reduce cost overruns and diminish the controversy of the past two decades on whether new systems were needed. Neither objective seems to have been accomplished to date. The group at OFPP was asked what problems still remained and how they could be solved.

> Along with Dave Baker, we talked with Chuck Clark, Marty Conley, and Bill Townsend at the Office of Federal Procurement Policy. Dave said that we would also want to talk to Fred Deitrick, who is the author of the A109 (government policy on major systems acquisition) document.[14] He's living in Florida and is now a consultant. He may be asked to rewrite A109. The problems are (1) it takes a long time to develop a major system and (2) the government should be more selective in its systems. To develop a major system takes ten, twelve,

fifteen years. Obviously we know that there is only a five-year plan and therefore this isn't sufficient. Lots of changes can happen in ten to fifteen years. DOD, in particular, seems to work in a shotgun approach. There is no real overall funneling and planning and no process of elimination of projects. It seems as if all projects go on a parallel basis and there's a continual battle among all these projects to get the funds. This produces the natural result of some not having enough funding and others running hot and cold. There seems to be inadequate overall project planning, and insufficient guidance relative to which system should be given the go-ahead for development.

There needs to be much better homework at the front end of the acquisition process. (This was the recommendation of the Procurement Commission in 1971.) The Packard Commission,[15] the Gary Hart book, *America Can Win*,[16] and just about every group that has studied procurement in recent years say that better homework at the front end is required. Get the services and agencies to identify *common* needs. Get them to work together!

Another need is to shorten the line of communication for the project manager. Project managers must continually brief people. There's an inordinate amount of time spent (about 60 to 70 percent of his or her time) justifying each project.

The government can save money by developing common systems. An example of the need for commonality in weapon systems is that the Air Force, the Marine Corps, and the Navy all are developing their own tactical air support systems instead of having one single system. Bill Townsend had previously been with the Cruise Missile Office, and he said that they had great success using commercial systems (off-the-shelf guidance systems, hardware, etc.) for that missile system; and that it was a very efficient and cost-effective program.

One problem is that technology changes. We have to be able to stabilize or freeze design with time and not continue to make basic changes in the fundamental design as programs progress unless the changes are pre-engineered into the system in a process known as phased, evolutionary development. This is an extremely important point!

On competition: "We may be competing things when it doesn't make sense. A study should be done to determine when competition should be used." An across-the-board implementation of competition in all cases doesn't make sense. This is one of the effects of the Competition in Contracting Act. It may have been a tremendous overkill and may have actually cost more money and more time in terms of securing items.

The executive branch is not asked for a feedback on laws, and many times Congress will pass a law not knowing what the real impact of that law will be. As a result, many times these laws accomplish the reverse of what was originally intended.

What about financial systems for tracking of procurement budgets and cost? It was said that DOD has a working group that has identified acquisition ADP (automatic data processing) systems and is trying to get a handle on this.

Joe Wright, the deputy director of OMB, made a speech relative to reforming the government's accounting system and Don Regan, as reported in the *Washington Times* on the 25th of August, 1986, made a statement that the government doesn't even *have* an accounting system.

A very important point brought up about project managers is that they should be informed that they will be protected and not punished, chastised, or fired for delivering bad news. In fact, they should be rewarded for bringing problems to the top that need to be resolved at an early phase rather than covering them up in the fear that the emergence of a problem will cause the program to be cancelled, the funding to be cut, or public or congressional support to be diminished.

Overall savings have not yet been proven as a result of legislatively increasing competition. On the contrary, the systems acquisition process itself is taking longer and costing more.

We also interviewed other people in the Office of Management and Budget. Bill McQuaid,[17] when asked about the government's financial management system, said it was "like a room full of worms":

> *DOD has identified 158 major automated procurement and finance systems in the Department. There is an objective to cut this by 50 percent in five years. They would like to get down to 130 systems by the end of 1986, 66 systems by the end of 1990. There is an attempt to try to get fewer financial management, payroll, and accounting systems within DOD. The presently existing number of accounting systems really would go into the thousands if you would go down to project level.*

Bill McQuaid and Dave Gribble[18] in OMB agreed that there was very little consistency in government financial management and accounting systems and that a tremendous amount of work has to be done to bring order out of chaos.

All of the above inputs and comments were duly noted and strongly considered in synthesis with literature surveys to develop the framework for action in the final chapter of this book.

Congress—And the General Accounting Office

We found that actions of the legislative branch of government have a profound if not overriding effect on the ways major systems and projects are acquired. According to the Constitution, "All legislative powers [are] . . . vested in a Congress of the United States" and "The executive power [is] . . . vested in a President." But because of the checks and balances built into our form of government, the Executive Branch often finds itself proposing legislation, and the Congress often finds itself in a management role. Because the Congress controls the money, Congress has become a key player in deciding which major systems are built and when. The Constitution also prescribes that the Congress will "raise and support armies, *but, no appropriation of money to that use shall be for a longer term than two years*" (emphasis added). As Chapter 3 shows, major programs and their phases far exceed two years in duration. The two-year restriction for defense systems seems to be a major hindrance in the efficient acquisition of major weapons systems since yo-yo funding has been cited as a huge cost driver, accounting for 30 to 50 percent cost growth in major systems.[19]

Congressional members have a very difficult and complex job because they must balance the desires of their constituents with the public welfare, and often the two goals do not agree. Their job is enormously complex, especially in this age of high technology. The matters they must consider are staggering both in variety and quantity. They must enact legislation in virtually every area of human affairs. Acquiring the major systems required to operate this country is only one of the myriad areas they must become informed on well enough to cast a vote. A large percentage of the persons in the Congress are from political, business, educational, or civic backgrounds and tend not to possess the detailed technical, engineering, and scientific expertise required to understand modern civil, defense, information, and scientific projects. So Congressional members must rely on credible experts both in the legislative and the executive branches of government to provide them with facts on which to base informed decisions. These facts must be weighed against a plethora of information and opinions of special interest groups to arrive at effective legislation. Often, as hinted in the interview with Dave Baker,[19] the Congress is not provided with sufficient in-depth analysis of the true potential impact of their proposed legislation. But they must vote a "yes" or "no" on bills of sweeping importance.

Political banter in Congress regarding major systems that was exposed by the following 1986 House Armed Services Committee news release[20] illustrates the Congressional dilemma:

> *WASHINGTON—Rep. William L. Dickinson (R-Ala.), ranking Republican on the House Armed Services Committee, today charged that many votes on the defense authorization bill this year "reek of hypocrisy."*
>
> *"No one should be surprised that pure politics is conducted on the House floor," Dickinson said. "But what I'm talking about isn't pure politics, it's pure hypocrisy."*
>
> *Dickinson cited inconsistent votes, with some Members "preening their fiscal responsibility feathers by voting to cut the dollar total in the bill, and then turning around to vote to add individual weapon systems to the bill."*
>
> *Dickinson listed several of the votes taken last week that prompted him to speak up.*
>
> *"First", he said, "Many of the same Members who voted* against *the President's Strategic Defense Initiative (SDI) by reducing its budget $2 billion below the President's request, several hours later attempted to stuff a Trident submarine into the defense bill. It's inconceivable to me that these arms control advocates could vote* against *SDI (a defensive system) and* for *a Trident, which carries 192 nuclear warheads." The Trident was not even supported in the Armed Services Committee. Other examples:*
>
> > •*Votes* for *the Midgetman missile, . . . but* against *tests that would be required to validate the performance of a warhead for that system. (Dickinson noted, 'this*

strategy could produce the ultimate liberal weapon system—one that was highly survivable but threatened no one, since it did not carry a warhead.')

•*Votes* for *the prohibition of chemical weapons, . . . but* against *removing aging chemical weapons from Europe, where they pose a significant hazard not only to the soldiers who must maintain and guard them but also to the civilian populations of our allies.*

•*Votes* for *requiring U.S. adherence to the SALT II treaty, . . . but* against *a Republican amendment to link U.S. compliance to that of the Soviets.*

•*Votes* for *procurement reform and lower defense budgets, . . . but* against *the production of weapon systems at efficient rates. (Dickinson pointed out that the House vote 'had reduced the quantity of F-15 aircraft from 48 to 24, thus increasing the unit cost of each aircraft by over $6 million.')*

"Particularly galling to Dickinson was the vote in the House to include the T-46 trainer aircraft in the defense budget. He said, 'Many of the very people who voted to cut the defense budget Friday led the fight to stuff the T-46 into the budget Monday. The T-46 is a $3 billion program of airborne pork—a program the Air Force wants to junk, a program that wasn't even included in the $320 billion budget request these same people call bloated. Dickinson noted that New York State Democrats voted 17-to-1 in favor of cutting the defense budget (with Rep. Samuel S. Stratton the solitary nay), while all supported adding money for the T-46 [which is built in New York by Fairchild].

In our interviews with project managers, administrators, and procurement officials in civil agencies, military services, and the space program, it became clear that many were pointing to Congress as the source of problems in the systems acquisition process. Indecision, slow or erratic funding, and conflicting requirements were cited as reasons for inefficient programs and enormous cost growth. Oddly enough (or perhaps not so oddly) when we moved into Congress, fingers were pointed back at the executive branch. One Congressional staffer, a member of the House Armed Services Committee staff, had the following inputs:

The Congressional staffer said that there has been a great debate about DOD expenditures since 1981. Since then, they have spent $2 trillion, and debate continues about the waste, fraud, and abuse in the Department of Defense. He said we should review the Packard Commission Report (which we had done). He said the blame should not all go on Congress but that the administrative branches are also at fault. For example, then-Secretary of Defense Caspar Weinberger deserved a lot of criticism because he constantly claimed that the United States was weak militarily and needed to get stronger: "No matter how high the defense budget is, he always wants more."

The DOD and military services want flexibility and they claim that Congress is micromanaging. They want the flexibility to do the job. Congress is saying (according to this staff member) that some of the answers involve baselining, milestoning and "streamlining" of project management.

The Congressional staffer said that the biggest thing that has helped is to have some budgetary constraints. He said Congress is prepared to have the discipline to meet a goal. He also said the Gramm-Rudman amendment and similar legislation has really helped even though a lot of people have complained about it because it forces the system to review each program in depth and make some tough decisions rather than just trying to get everything funded.

He said that Congress has criticized the high cost of spare parts and the large number of undefinitized contracts, that the DOD spends about $140 billion a year in procurement, and that *"we're just simply spending money rather than getting jobs done."*

When we asked this staff member of the House Armed Services Committee what the real problem was he said that *the project manager really doesn't manage his project.* He really just mediates. "He's being driven by the system."

The staffer spoke of the huge unobligated balances that are showing up in the DOD and how we budget for inflation. He alluded to the missing $10 billion (and it could be between $30 and $40 billion) in the DOD that was allocated for inflation that never materialized. He said one of the principal problems is that the *budget input process is not connected with the output accounting process.* In other words, we don't know what we've spent against the budget. He said "Congress can't track where the money went—the process is so complicated that you just can't do it."

He said the problem in the DOD is (although you won't find this written down in any of thousands of documents) that you're measured by how effectively you get dollars from the Congress and how much you spend—*not how much you save.* This becomes a self-fulfilling prophecy because when you try to spend more money, of course, you *do* spend more money. And, if you don't obligate all your money, then they (the layers above the project manager) say "you're behind schedule" and "this will hurt us in getting next year's appropriation." The real question should be, "How cost-effectively do you spend the money and how much money can you *give back to us* at the end of the fiscal year?" He admitted that good programs *do* exist.

We then asked him this question: "Do you think that the current round of proposals is really going to basically change our way of doing business or will we just get more of the same? In other words, is the situation hopeless under our current form of government or can something be done?" Other questions that were asked (but not answered) were "Has legislation more often than not created exactly the opposite effect to that desired? Could there be more in-depth planning to determine the long-term impact of implementation before hastily enacting more band-aid legislation?"

The staff member said that they (DOD) would rather put up with restrictive language than back up on getting the money. He said getting the money is the key objective: "As long as we keep giving them the same amount of money (or more), then they are happy."

About our form of government, he said it wasn't necessarily supposed to be efficient, but it is there to protect the individual's rights. He said the problem is that we can never get a consensus; therefore we are at risk to a potentially well prepared adversary. He said, "Our security may be based more on our ability to vote than our ability to efficiently procure."

At this point he recommended a book, *How Democracies Perish,* published in 1983 by Harper & Row.[21] (It might be more aptly titled *Look Out for Those Rus-*

> *sians!* It presents some sound warnings against complacency.) He said another problem is that *Defense is not policy oriented but procurement oriented*. And there are some basic questions, such as: Will defensive deterrents replace offensive weapons? These questions must be answered clearly before a specific plan can be worked out.
>
> He said we must develop the discipline to make the policy decisions on hardware *up front* rather than adopting the policy of giving a little bit to Rockwell, a little bit to McDonnell Douglas, a little to Boeing, and so on. He said no one has been willing to make the trade-off—not the services, not the DOD. He said more input should be provided from the field commander, that is, the user advocate (such as NASA does with the astronauts).
>
> We asked what his overall recommendations were. He said, "Well, Gramm-Rudman may be a blessing—it forces discipline; it causes trade-offs and gives us tough trade-offs to make." He added, "The most powerful leader is the dollars, not the policy."

Despite the tone of the comments from this interview, and from what we know about our Constitution, our form of government, and the conflicting pressures on legislators, it is clear that the Congress cannot escape responsibility for the situation as it is. The thought that freedom and efficiency are incompatible does not play well with the current efforts to increase efficiency. The fact that there have been successes that everyone is trying to emulate indicates that most people believe that something can be done to make the process more efficient without jeopardizing our freedom. Since Congress, in general, makes the laws and the President carries them out, the ball always comes back into the court of Congress. But we must remember that we, the taxpayers and voters, are also in the same court. Our Congressmen and Congresswomen reflect what they think *we* want. Could it be that we ourselves (as Pogo implied) are the problem? Let's not rule this out now as we continue to explore what can be done about turning our government into a highly efficient mechanism for providing the large, high-tech projects needed to fill our civil, defense, and information needs.

One frequently overlooked but ever-present source of information on how to more effectively manage large government projects is the General Accounting Office (GAO), an arm of the Congress. This $350 million per year office has the following as its basic purposes[22]:

> *to assist the Congress, its committees, and its members in carrying out their legislative and oversight responsibilities, consistent with its role as an independent nonpolitical agency in the legislative branch; to carry out legal, accounting, and claims settlement functions with respect to the Federal government programs and operations as assigned by the Congress; and to make recommendations designed to provide for more efficient and effective government operations.*

GAO has produced hundreds of reports that identify areas where greater efficiencies can be achieved in the big projects of government. The problem is that Congress does not always take action on the recommendations in these reports, and they are frequently treated by the agencies and departments of the executive branch more as a harassment than constructive criticism. Further, GAO has the job of establishing the form of government accounting systems but has no enforcement tool, other than Congressional legislation, to have these systems adopted.

Part of the reason that GAO reports are not always given great attention, or are ignored entirely, is the basically good but confusing practice of permitting the target office or agency to provide rebuttals. Rebuttals are often bound into the report itself. Rebuttals are sometimes responded to by GAO, and then again by the agency reviewed. This written debate cycle that makes up many GAO reports often causes the initial issue to be obscured in a sea of words, charges, and countercharges; and the reader often has little clue as to what the most effective final action, if any, should be. The lack of action on these reports is unfortunate, especially considering the large effort spent to produce them. And the GAO reports are high-quality, professional, and diligently prepared documents.

Since the GAO is nonpolitical and is headed by a Comptroller General appointed by the President with the advice and consent of the Senate for a term of 15 years (which is often renewed), the office is an ideal source of sound, independent advice. Greater use of and attention to GAO findings on major systems acquisition would result in the more effective use of a highly qualified and professional service that is already available. In addition, greater emphasis on adhering to GAO recommendations would complement rather than duplicate activities of the plethora of citizens' lobbying groups such as the National Taxpayers Union, Common Cause, the Coalition for Common Sense in Government Procurement, the Project on Military Procurement, and various whistleblowers' protection groups. The GAO produces many acceptable recommendations on how to increase efficiency in major systems acquisition. But in order to reap benefits from the GAO's painstaking and meticulous work, Congress must be receptive to its constructive suggestions, and agencies should be more receptive to its constructive criticism. The point here, again, is that the effective use of an *existing organization* can help avoid the establishment of more commissions, study groups, committees, blue-ribbon panels, and citizens' lobbying groups—and the offices, agencies, legislation, regulations, and policies that *they* may recommend for the care, feeding, and "oversight" of the systems acquisition process.

Megasavings in Megasystems

In the past, it has been difficult for the average citizen to fathom the amount of money required to develop and procure our large systems. This is, perhaps, part of the reason so much attention was directed during the mid-1980s to the costs of spare parts, rather than the major systems they support. It is much easier for the average citizen to be concerned over the cost of a $14,000 copilot's chair (base not included) for the C-5 cargo aircraft[23] or to become appalled at the costs of a $44 light bulb, a $2,043 hex nut, or a $387 flat washer[23] than to question the multimillion and multibillion dollar costs of the major systems that use these parts. If dramatic savings can be achieved in spare parts procurement, as was done in the mid-1980s, then should we not expect similar savings in megasystems?

Just how much money are we spending and how much money could we save in major systems acquisition? Are we talking big bucks? And if we saved this money, could it really be returned to us in the form of lower taxes? Or could it result in filling more of our needs at the same cost?

Current budget year and approximate run-out figures for many major systems are now available, and these give a general view of the tremendous size of present and future planned major systems expenditures. For example, in the estimated fiscal year 1987 defense budget of $320 billion, over $100 billion was estimated as procurement.[24] In its report on "Program Acquisition Costs by Weapons System,"[25] the Department of Defense lists projects totaling $64.122 billion. Thus, about 65 percent of the estimated defense procurement budget was for the acquisition of major systems. This number does not include the millions of additional dollars required for training, operation, and logistics to support the major systems; huge capital facility expenditures for much of the "brick and mortar"; and equipment needed to research, build, operate, and dispose of defense and nuclear projects. Nor does the weapons procurement budget figure include the many multimillion dollar communications, information, data, and *management systems* being procured to help manage the Department of Defense and the other Federal agencies. Also included in the category of major systems are the numerous transportation systems, communications systems, space systems, and civil projects that are either wholly supported or subsidized by the federal government. Even with the conservative assumption that 25 percent of the total federal budget for one year is attributable to the development, acquisition, operation, support, and disposal of multiyear, high-tech complex major systems; the one-year expenditures could approach $250 billion. With all its zeros and decimals, this number can be expressed as $250,000,000,000.00 which is a whole lot of money. When the *run-out* or *life-cycle-costs* of all of the major systems, present and future, are considered, the number could reach ten to fifteen times this amount.

When one looks at the projected costs of individual systems, one can see why this total number is so large. For example, the Army's Bradley Fighting Vehicle: The Army was scheduled to purchase 870 vehicles in fiscal year 1987 for a price of $1.142 billion, which amounts to about $1.3 million per vehicle. This is cheap, however, compared to the M1 Abrams Tank, which at roughly the same production rate, cost $2.375 million each. These land vehicles, however, are bargains compared to some of their airborne and seagoing counterparts. The Air Force's F-16 Falcon fighter aircraft runs 16 million dollars a copy—at a production rate of 216 per year. Imagine a single pilot being responsible for flying a $16 million aircraft in combat! And these are all combat vehicles—subject to being damaged, marooned, or shot down during a conflict. The list of multimegadollar systems goes on. For 75 percent more than the price of an F-16 we bought F/A-18 single-seat Hornet aircraft for the Navy at $28 million each (quantity of 120 in fiscal year 1987).

Orders of magnitude more costly are the ships and submarines we are buying for the Navy. To reach the goal of a 567-ship Navy, and on to the ultimate goal of a 600-ship Navy, the United States is spending $583 million each for four nuclear attack submarines; $843 million each for three guided-missile–carrying destroyers; one billion dollars each for two guided missile armed cruisers; and $1.5 billion for a single Trident submarine (not including missiles at $68 million each). All of the systems mentioned so far are "nonexpendable"; that is, they should be good for many missions. But there are families of *expendable* weapons that also reflect phenomenal costs, and they are the missiles, the high-tech artillery of modern warfare. Even the smallest missiles such as those used in the Army's Multiple Launch Rocket System (MLRS), and Stinger systems cost as much as a new subcompact car per set ($6,500–$7,000) in production quantities of several to tens of thousands per year. Many missiles cost over $100,000 each—about the cost of the average home (examples are Harm, Maverick, Sidearm, and Sparrow. Larger missiles cost over *$1 million per copy* (Patriot, Harpoon, Phoenix, Ram, Tomahawk, Advanced Medium Range Air-to-Air Missile, and Ground-Launched Cruise Missile). Two notable multimillion-dollar missiles are the Navy's Trident II and the Air Force Peacekeeper, or MX missile.

These are merely examples of some of the larger and more visible military weapons systems. (For more information on what major weapons systems are bought and how they are selected, read *How Many Guns are They Buying for Your Butter*, by Tobias, Goudinoff, Leader, and Leader.[26] But the government also spends billions of dollars on other types of systems that are used not for defense but to operate the government itself and to "promote the general welfare." Included in these are transportation systems (highways, railroads, canals, bridges, airways, and rapid transit systems), communications systems, information systems, management systems, space systems, postal systems, weather satellites, and many more. Examples follow:

- The National Oceanic and Atmospheric Administration's $1 billion annual budget contains funds for weather satellites costing $37.3 million each for use in global weather forecasting.
- NASA is planning to spend $18 billion (not including launch costs and

experiment costs) on a space station and orbital laboratory to be placed in space during the next decade.

- The U.S. Air Force spends about $3 billion a year on mission-critical software for their major systems.
- The Social Security Administration is spending over $800 million in a five-year effort to modernize their data handling and communications system.
- The U.S. Patent and Trademark Office is developing and installing an Automated Patent System that will permit their 4,000 or more users per week to easily and quickly search their almost 30 million records. The information system will cost over $550 million.
- The Federal Aviation Administration is spending over $197 million to modernize the nation's computerized air traffic control system.
- The U.S. Army is planning a $670 million computer system to automate logistics and reserve personnel information. The system would be used, among other things, for calling up the reserves in the event of a national emergency.
- The U.S. government is modernizing and extending its federal telecommunications system called FTS 2000. The estimated cost over its ten-year development and installation time is in excess of $4 billion.
- The Department of Defense is spending $2 billion to modernize its Worldwide Military Command and Control System.

The programs listed above, in *numbers* of major programs and systems, barely cover the tip of the iceberg. There are thousands of civil and military construction projects to be carried out by the U.S. Army Corps of Engineers, the services, and other agencies of the federal government, which extend well into the 1990s and beyond. Added to the information systems category are automated systems to assist in the acquisition process itself. The portion of the budget that will go toward automated systems of all types is expected to grow dramatically in the next decade.

The total potential savings in all civil and military systems through more efficient acquisition are enormous. A study titled "U.S. Defense Acquisition: A Process in Trouble,"[27] completed in March 1987 by the Center for Strategic and International Studies at Georgetown University, estimated that up to $50 billion a year could be saved in defense programs alone through (1) reduction of oversight, auditing, and regulations; (2) elimination of program instability and poor estimates; and (3) avoidance of financial errors, requirements excesses, and overspecification. These savings equal about 25 percent of the major systems acquisition costs of the Department of Defense each year. Assuming that like savings could be achieved for the major programs and projects of all of the other agencies, cost reductions of up to $100 billion per year could be achieved. These savings could go a long way toward elimination of the federal deficit and reduction of the national debt.

As we move into future weapons systems such as advanced tactical fighters, stealth bombers, stealth fighters, advanced electronic countermeasures, the aerospace plane, and the satellites, detectors, and battle management systems of the "Star Wars" or Strategic Defense Initiative (SDI) programs, systems will become even more complex and costly. In the future, the opportunities for cost savings and cost avoidances through stabilized funding, meticulous financial tracking, excellent management, and well-thought-out procurement policies will be even greater than they are today.

From our interviews and discussions with individuals in government and industry, it became clear that several routes were available to bring about sizeable savings in expenditures for major systems. The first, most important, and most often repeated recommendation was that the *funding process be stabilized*. Some mechanism or philosophy is required that will permit full funding of major system acquisition phases. By providing stable funding, huge cost avoidances are possible. Stable funding avoids the enormous start-up and shut-down costs associated with erratic funding and minimizes program cost growth. A second, very important method of saving dollars in major systems was found to be the use of common systems among multiple users. An example of the immense cost savings made possible in multi-user systems is in software and computer systems. Industry and government alike have found that off-the-shelf, commercially available computer software and hardware systems can be acquired and maintained at a small fraction of the cost of customized systems. Spreading the development and production costs over many users dramatically reduces the cost to any given end user. Standardized or generic designs, products or projects that can be adapted to many uses are *orders of magnitude more cost-effective* than specially developed systems for unique applications. Other areas—such as stability of management, elimination of counterproductive and complex acquisition methods, and using existing organizations and expertise rather than duplicating them—are equally likely to produce sizeable savings in the total major systems acquisition budget, as well as in *other* areas of government.

People, Politics, and Procurement

It is evident that no single lobbying group, government agency, or politician will be able to instigate the long-term, far-reaching changes needed to produce in-depth, meaningful changes that will eliminate the root causes of the problems in major systems acquisition. Demand for reform in the systems acquisition process is coming from multiple sources. Industry is desperate for clearer guidelines about what to expect in the future. Those trapped within bureaucratic mazes in the government itself are becoming exceedingly frustrated in their inability to get things done. And the private citizen is tired of hearing alarming stories that have returned each year just as surely as the swallows have returned to Capistrano. However, the wheels of progress are turning toward financial stewardship in government. Balancing the budget has become a national priority. And steps are already being taken to start eradicating the gigantic national debt. High-tech communications, planning, information management, and financial tools are becoming available and will help manage the enormously complex projects that are now being developed and produced—and those that will be developed and produced in the future.

Professional societies, nonprofit groups, trade organizations, research institutes, universities, foundations, and citizens groups are marshalling their best experts to work in the public interest toward efficiency, economy, and effectiveness in government. Answers are becoming available about how to cooperatively promote excellence in government. We have watched excellent industries do it; we have watched other nations do it; we have hired some of the best advisors in the world, many as government contractors, to tell us how to do it.

As John Naisbitt pointed out near the beginning of this decade in his book *Megatrends*,[28] we have moved from an industrial-centered society to an information-centered society, from forced technology to high technology, from a national economy to a world economy, and from short-term goals and planning to long-term goals and planning. These profound changes are being felt in the conduct of business in the largest single organization in the world, the U.S. government, and it will be felt heavily in the acquisition of its biggest projects. In this multiple-choice, multiple-option society, we must *use* the newly born high technology to develop and manage these systems rather than let the growth of high technology control us. We must use it to run our own country's largest projects. Our participatory democracy permits each of us to have a louder voice then we often realize in how our government buys its big systems, and our willingness and ability to exercise this voice is improving because we are becoming better informed through channels provided by the high-tech society itself. Knowledge that we have this voice, and the willingness to put it to work, will have a pronounced effect on excellence in planning, scheduling, estimating, developing, building, and operating our major systems.

In another book by Naisbitt,[29] he and his co-author point out that the corporation in the United States is being "reinvented." Our government need not be reinvented; it need merely be rediscovered as an implement with which to govern by the consent of the governed. All of the tools are available: The organizations are established; the mechanisms are in place; the good people are on board; the facilities are on line; and the industrial, research, and university complexes required to support it have been mobilized. These unprecedented resources can now be put to use in one of the largest tasks ever undertaken, the organization, national planning, financing, and management of our present and future major civil and defense systems inventory. As will be pointed out in depth in Chapter 7, members of the Congress have never really been given a comprehensive, credible, and viable alternative to merely "voting their constituencies" relative to major systems and projects. We believe that their participation in and continued approval of a dynamic plan constitutes this desperately needed and viable alternative.

Endnotes

1. John C. Livingston, and Robert G. Thompson: *The Consent of the Governed*, Macmillan, New York, 1963.
2. Benjamin S. Blanchard and Wolter J. Fabrycky: *Systems Engineering and Analysis*, Prentice-Hall Inc., Englewood Cliffs, NJ, 1981.
3. Harry A. Seymour: "United States Navy," *World Book Encyclopedia*, Field Enterprises Educational Corporation, Chicago, 1967.
4. Interview with Richard A. Reynolds, Director of Defense Sciences: Defense Advanced Research Project Agency (DARPA), Washington, DC, September 15, 1986.
5. General Accounting Office: *Listing of Recent Defense Acquisition Studies Bibliography*, General Accounting Office, Washington, DC, 1986.
6. Commission on Government Procurement: *Summary of the Report of the Commission on Government Procurement*, Commission on Government Procurement, Washington, DC, December 1972.
7. Rodney D. Stewart served as chairman for the final year of the study group.
8. Commission on Government Procurement: *Major Systems Acquisition—Final Report*, Commission on Government Procurement, Washington, DC, January 1972.
9. Commission on Government Procurement: *Special Report on Acquisition of Major Systems—Draft*, Commission on Government Procurement, Washington, DC, December 1972.
10. Thomas J. Peters and Robert H. Waterman, Jr: *In Search of Excellence*, Harper & Row, New York, 1982.
11. General Accounting Office: *Advantages and Disadvantages of Centralized Civilian Acquisition Agency*, General Accounting Office, Washington, DC, November 1986, NSIAD-87-36.
12. Interview with James Gregory, Director of Procurement (retired), Strategic Defense Command, Huntsville, AL, August 1, 1986.
13. *Government Executive*, National Journal, Inc., Washington, DC, March 1987, Vol. 19, No. 3.
14. Office of Federal Procurement Policy: *Major System Acquisitions: Discussion of OMB Circular No. A-109*, Office of Federal Procurement Policy, Washington, DC, August 1976, OFPP No. 1.

15. Packard Commission: *A Quest for Excellence: Final Report of Packard Commission, 2 vols.*, Packard Commission, Washington, DC, June 1986.
16. Gary Hart and William S. Lind: *America Can Win,* Adler & Adler, Bethesda, MD, 1986.
17. Interview with Bill McQuaid: Office of Management and Budget, Washington, DC, August 25, 1986.
18. Interview with David Gribble: Office of Management and Budget, Washington, DC, August 25, 1986.
19. Interview with Dave Baker, Acting Administrator: Office of Federal Procurement Policy, Washington, DC, August 25, 1986.
20. House Armed Services Committee: *Dickinson Says Floor Votes Reek of Hypocrisy—News Release,* House Armed Services Committee, Washington, DC, August 19, 1986.
21. Jean-François Revel: *How Democracies Perish,* Harper & Row, New York, 1983.
22. *United States Government Manual 1987/1988,* General Services Administration, Washington, DC, 1987.
23. Christopher Cerf and Henry Beard: *The Pentagon Catalog,* Workman Publishing, New York, 1986.
24. Interview with Al Berman: Office of Management and Budget, Washington, DC, August 25, 1986.
25. Department of Defense: *Program Acquisition Costs by Weapon System,* National Technical Information Service, Springfield, VA, February 1986, PB86-153004.
26. Sheila Tobias, Peter Gordinoff, Stefan Leader, and Shelah Leader, *What Kinds of Guns are They Buying for Your Butter?* William Morrow, New York, 1982.
27. Center for Strategic and International Studies: *U.S. Defense Acquisition: A Process in Trouble,* Georgetown University, Washington, DC, March 1987.
28. John Naisbitt: *Megatrends,* Warner Books, New York, 1982.
29. John Naisbitt and Patricia Aburdene: *Re-inventing the Corporation,* Warner Books, New York, 1985.

CHAPTER TWO

The Anatomy of a Successful Project

Unfortunately, there are few absolute quantitative standards that can be used to determine whether a large program was a success or a failure, only qualitative judgments by the program's supporters or critics. Success in the marketplace is measured quantitatively by determining sales, profit, market share, and the like. Megaprograms have generally been judged by programmatic measures (cost, schedule, and performance). Obviously it is difficult to classify a program as a success if there are large cost overruns, schedule slippages, or performance deficiencies, or as a total failure if there are minor changes and small cost or schedule increases. Even the most successful program has some problems, and even the least successful programs have some good points. But no one, so far, has developed precise measurements that would permit classification of all programs against the three criteria of cost, schedule, and performance. To identify successful programs we must go to those which it is generally agreed have accomplished their objectives and have done so within close boundaries of their originally advertised resource and time constraints.

What magic mixture of ingredients does it take to plan, estimate, and carry out a megaproject that produces its desired effect on schedule within reasonable cost boundaries? How is the right combination of methods, techniques, tools, people, organizations, and ideas selected, blended, distilled, and formed to create success? And what is the proper sequencing of activities required to synthesize the diverse facets of a large high-tech project and bring it to a successful conclusion?

The answers to these questions can be found, in part, by closely observing those projects that have been generally recognized as successful. In virtually every arm of the government, and in the private sector as well, there are examples of outstandingly successful projects. We can look at selected outstanding megaprojects to determine if there are traits that can be emulated in future projects to ensure or enhance their success. Another body of information is resident in big projects that were *not* so successful.

Can we discover and eliminate the causes of failure? Still another source is the body of information available from research institutes, committees, blue-ribbon panels, universities, and industry/government studies. A critical review of these studies will reveal factors that are likely to result in success or failure.

History will surely reflect that one of the most outstanding, if not the most spectacular, megaprojects of this decade was the launching of men to the moon and returning them safely to earth. The giant rockets used in this highly successful project had their roots in another culture and another country. The German scientists that worked on the V-2 rocket during World War II were familiar with failure. Of the 3,600 rockets launched against Britain and targets on the Continent, some 25 percent failed largely because of premature detonation (airburst).[1] But early rocket development in Germany built valuable experience for the rocket team, which was later transferred to the United States to test refurbished V-2s, precursors to our modern missiles and space vehicles, at White Sands Proving Ground.

One of the key scientists who helped in the transfer of rocket propulsion technology to the United States was Dr. Eberhard F. M. Rees, deputy to Dr. Wernher Von Braun and, later, Director of the George C. Marshall Space Flight Center in Huntsville, Alabama. We interviewed Dr. Rees in his home atop Monte Sano in Huntsville.[2] We asked him to tell us of successful (and unsuccessful) programs with which he has been associated and what lessons he has learned from his experiences in high-technology rocket systems development since arriving in this country in 1945. This interview, recounted in its entirety in Appendix A, provides some revealing insights into causes for success and failure.

It is apparent from the interview that Dr. Rees, as well as many others who have been involved in highly successful projects, is a firm believer in in-depth involvement by government technical personnel. Government-technical personnel are able to bring greater technical expertise to bear on the project because they have access to the independent internal expertise of government laboratories which can be put to work on specific problems. These government engineers are less likely to be driven by the profit motive which could cause short-cutting if the contractor were left to his own devices. They can more easily bring the broad experience from other programs to bear on the problems at hand, and, provided that their own management is not overly driven by pressures to meet schedules despite technical problems, they are often able to give a more objective look at the technical needs of programs. NASA personnel were cut back prior to the space shuttle program which caused less hands-on involvement of government personnel in contractor plants, and both government and contractor decisions may have been flavored by pressures to meet a launch schedule. The detrimental effect of fewer monitoring government personnel, is addressed in the interview with Dr. Rees. The Presidential Commission on the space shuttle *Challenger* accident cited schedule pressures in their report[3] which said that "the prevalent attitude in the program appeared to be that the shuttle should be ready to emerge from the development stage, and managers were determined to prove it 'operational.' "

After the very successful manned lunar exploration program, broad Congressional and public support for the space program gradually faded. Priorities were directed elsewhere due to international and domestic problems. The diminishing emphasis on a "presence in space" by our leaders, only recently replaced by a growing enthusiasm for space exploration, resulted in lower funding, lower personnel ceilings, and possible missed opportunities to equal or surpass the Russians in their space technology.

Although the Apollo program had military beginnings, it was a civilian effort. The moon-landing and return project was given a high national priority by the President himself. The whole nation was behind it. It captured the imagination of virtually every citizen, from scientist to schoolchild. Only a very broad goal was provided by the President: to land a man on the moon, and return him safely to earth, by the end of the decade. The Congress was behind it. The funds were set aside to do the job within the broad schedule of "by the end of this decade" (the 1960s). The German-born scientists and their American rocket engineer counterparts who were instrumental in making the proposal for this mammoth project were already in place at Huntsville, Alabama. And new centers were established at Houston, Texas, and Cape Canaveral, Florida. Recalling this setting, what did we learn from the interview?

First, Dr. Rees and Dr. Von Braun were always available to their senior engineers and scientists for the open discussion of technical or administrative problems as the important propulsion systems for the Apollo program were being developed. Despite what one might think based on their previous background and their work for the regime in Germany, where authority seemed never to be questioned, and in spite of a potential language barrier, these adopted American scientists formed a remarkably democratic, and entrepreneurial team. Ideas were welcomed. Open discussion of problems was encouraged. The many American-born engineers, scientists, administrators, technicians, and workers who joined the group as it expanded from the tiny band of immigrant Germans were readily embraced into the team. We remember that Dr. Von Braun's door was always open; he quickly grasped ideas and suggestions. He was a great leader and communicator. He, Dr. Rees, and the other German-born scientists adapted amazingly well to the language, the way of life, and to their new work environment. There were few formal barriers to communication, and a participatory style of management was quickly adopted. Detailed, dynamic planning was carried out early in the program with the participation of government and industry personnel, and systems requirements and definitions were baselined early in the program.

Second, Dr. Rees and his co-scientists always listened to the ideas presented by the industrial firms that supported the space program. But they were constantly aware of how the profit motive can sometimes adversely influence technical decisions. Once the ideas, inputs, cost estimates, and impacts were brought to the forefront and evaluated, decisions were made based on what was best for the program and not what produced the most profit. The primary objective of the Apollo program was to place a man on the moon and not to create an industrial base, but the formation of an industrial and technology base to support the space program was one of the derived benefits.

A third observation about the development of rockets for the Apollo program was that the development team believed in retaining an excellent in-house capability. Laboratories were formed to maintain expertise in every discipline—design, manufacturing, testing, electronics, structures, propulsion, and project management. And engineers or technicians from this home team were sent to contractors' plants to provide advice on a day-to-day basis. Initially, as Dr. Rees pointed out in the interview, the private firms intensely disliked this idea: But as the on-site personnel, and project office personnel who traveled to and from the contractor plants, began to demonstrate their technical knowledge and wield influence in the programs, the host firms began to see the substantial advantages of their presence. It appeared that our governmental system of checks and balances was also being applied to the customer-supplier relationship in an immense high-tech project: the Apollo moon-landing program.

Fourth, the team had management stability and continuity. The same group, with very little turnover, stayed with the program from the V-2 testing days at White Sands, New Mexico, through the latter stages of the Apollo program and into the early phases of the space shuttle project. The same engineer who directed rocket engine turbopump design for the V-2 rocket in Germany, for example, retained that responsibility throughout the evolutionary development of propulsion systems that led to the huge fuel pumps used in the Apollo moon rocket. Even though detailed design was done by contractors, the government engineer monitored progress and had a significant say-so in major engineering decisions. The propulsion systems leading to Apollo were excellent examples of the beneficial results of *life-cycle management* using a highly competent technical team.

Fifth, in the rocket developments leading up to the Apollo moon rockets, the combined benefits of phased acquisition and evolutionary development were evident. Development was approached as a methodical and sequential process. Detail of design definition was increased in each phase, and larger rockets benefited from the experiences encountered in the development of their smaller predecessors. Selection of the best contractors resulted from a narrowing-down process through each phase until the final production contractor was chosen. Phase A, or preliminary conceptual studies, was done in-house, occasionally with contractor support. Phase B, preliminary design studies, was opened to competition with two or three study contractors. The best of the study contractors was chosen to move into Phase C, development, and Phase D, production.

Pervading the Apollo program was a stability and continuity of support that should be given to all major Federal undertakings. Mr. Clarence Milbourn, Director of Contract Pricing and Finance for NASA, in a presentation to the first annual symposium of the Institute of Cost Analysis,[4] said, "We had full White House and Congressional support on the Apollo program. The nation seemed to be with us. We brought that program home pretty much at the expected cost. [It was] a stable program with stable budgets."

As mentioned earlier, even the most successful programs have some failures. The two notable failures so far in the space program were the Apollo spacecraft fire on January 27, 1967, that resulted in the death of three astronauts, and the space shuttle *Challenger* accident nine years later in which seven crew members died. It is interesting to note that both of these accidents could have been avoided if management had more carefully listened to those responsible for engineering and safety. Safety engineers had warned against using an almost pure oxygen environment in the Apollo spacecraft instead of air which is less than one-fourth oxygen. Technicians at NASA and Morton-Thiokol warned repeatedly that the solid rocket booster joint deficiencies in the space shuttle could result in "losing a flight."[3] Failure to recognize and heed internal warnings was a major contributor to both accidents. Dr. Rees' advice to "always listen" bears out the importance of management attention to detail and underscores one of the principal criteria for a successful program—careful and meticulous surfacing and elevation of serious engineering concerns to the highest levels of management, and the subsequent resolution of these critical problems.

Excellence in Military Projects

As mentioned earlier, the major weapons systems acquisition process in the Department of Defense has undergone searching and in-depth analysis by many organizations over the past several decades. In-depth reports containing recommendations have been prepared by the Rand Corporation,[5] the Packard Commission,[6] the General Accounting Office,[7] the Office of Management and Budget,[8] the Department of Defense,[9] the Commission on Government Procurement,[10] and the Congress[11] to name a few.

To provide a recent check point on the results of these studies, and gain more detailed information on a real project, we decided to select, for further study, the Army's Multiple Launch Rocket System (MLRS).

It is a common occurrence for a weapon system to be applauded by its own management as successful, but rarely does a service single out a weapon from among hundreds in its arsenal and highlight it as outstandingly successful; and rarely does the General Accounting Office (GAO) produce a glowingly positive report on a weapon system project. All of this has happened in the case of the MLRS project. The U.S. Army considers the MLRS an important example of a weapon that works. It had no cost overruns. The GAO gave the project a clean bill of health, which it rarely does for a program. The GAO usually investigates only those projects that are in trouble, but a 1982 GAO report *The Army's Multiple Launch Rocket System Is Progressing Well and Merits Support*[12] and a 1985 GAO report indicated that the MLRS "has excellent potential for significantly increasing the Army's artillery capability" and "has done well and merits continued backing by the Secretary of Defense and the Congress." The *Army Times*[13] stated, "The MLRS was well managed from the beginning. That is a major reason for its success."

The Multiple Launch Rocket System is an unguided, multiple launch, surface-to-surface battlefield missile defense system mounted on an agile tracked vehicle. It is available for rapid artillery counterfire and air defense suppression. Specially designed for use during surge periods when attacking forces present targets in sufficient quantities and densities to strain the capacity of available fire support systems, it is designed to complement rather than replace current fire support weapons. A $6.5 billion project including development and production, it is a conventional rather than a nuclear weapon.

Larry Seggel is the Deputy Project Manager of the MLRS Project. He is a big man,—articulate, outspoken, and gentle—hard-nosed in an amiable sort of way. Larry has been in the project since its inception and is a true product champion of the MLRS. In our interview with Larry,[14] we sensed that even though he was *Deputy* Project Manager, he really ran the show. His comments readily reveal some of the principles that have made the MLRS project outstanding. (Our emphasis is added to Larry's statements.)

RDS: *Why was the MLRS project so successful?*

MR. SEGGEL: *First off, we did a lot of planning before the MLRS even got started—a lot more than even now seems allowed. We tend to hurry like heck getting something started and then spend a lot of time sorting out why we should have done it some other way. Fortunately this program was allowed to be planned essentially by the project office. We got very little planning help from "20-minute planners." Maybe that had to do with the fact that we just didn't take kindly to a whole lot of help. But we did planning in-house and then we did a lot of work on strategy making sure that the signals we gave in our RFP (Request for Proposal), the things that we had in our contract, leveraged the contractor in a way that would cause the contractor to do and act and think how we wanted him to in order to make this program be what it was supposed to be.*

RDS: *Did you use incentive contracts?*

MR. SEGGEL: *Well, in the initial development stage, what is now called advanced development, in our program was called validation stage. We used competitive contracts. We had two contractors, LTV and Boeing, working to the same specs, going head-and-head to a shoot-off. In addition to the shoot-off, proposals were for the completion of the development and four years of production so that when he [contractor] came to us with his DTUPC [Design to Unit Production Costs] projection at the end of the validation stage, and says, "Boy, guys, this is how good I'm going to be," the biggest liar wouldn't win because he's going to have to sign a contract with three years of options on production at that value. So we spent a lot of time (it was about six months—probably six or eight of us) planning and wordsmithing.* Everything we wrote down in our RFP, we'd argue about and study to see whether it was going to be effective or make sure you got the proper reaction from eight different folks. If I said this, what does that mean to you? What would you do? And we kept working it that way and as a result (as a fundamental) we wanted leverage at all times in the contract in our relationships with our contractors. Any time there was a decision, any time there was a contract, we wanted the leverage on our side assuring us that the biggest liar couldn't win. We had financial leverage always. *After we got into production, it was clear to him that we were going to be establishing a second source.* We would establish a second source so that he'd never get a chance to run free. *Ultimately we didn't do that. We signed a multiyear contract which wound up being a far better deal because the contractor we chose, LTV, did a superb job.*

RDS: *You said multiyear. How many years was it?*

MR. SEGGEL: *Five years with two years of options. But LTV did an excellent job of performing to their DTUPC. I'd say without any qualms that they deserved it. They did a helluva good job and they're still doing a helluva good job on cost.* So I think it was the detailed planning and massaging and the great care in laying out the RFP and the actions in preparation for going into contract; the whole strategy of acquisition was the number one thing that set the ground work for everything else.

RDS: *Larry, in this planning process, did you get inputs from field type guys that would have to use this system?*

MR. SEGGEL: *Well, we worked together all the way. We worked with our user advocate at Fort Sill during the whole of what we call the concept definition stage—those guys were members of our team. Any time we'd work up a strategy that particularly had to do with interfaces regarding requirements or operations, we would bounce that business off them. Now in terms of business aspects of acquisition, they had very little impact because that's not their bag. That's what I'm good at;* they're good at soldiering. I'm good at managing and developing. *As long as we talk together we usually turn out something pretty good.*

During our interview with Larry, one of the other team members, Richard Paladino, was present. It was obvious that Larry's enthusiasm for the project had infused Rich. The MLRS project placed heavy emphasis on involving the end user in development. In this case, the using troops were represented by an organization called TRADOC (Training and Doctrine Command) at Fort Sill, Oklahoma. Some clues to the success of MLRS can be seen in the relationship of the project office to the user and the resistance to adoption of "gold plating." We resume the interview with a comment made by Rich:

MR. PALADINO: *I think Larry can explain that we were doing the planning and acquisition strategy and that sort of thing here, but I think one of the other key successes was [that] the TRADOC side of the house was defining the requirements of the system so that they were unchanged, in fact, throughout the development.*

RDS: *Was this the key reason for success, or were there other reasons?*

MR. SEGGEL: *I'm coming to that.* We're coming to the real reasons for success. *Those guys at Fort Sill and our project managers and the teams that work for each one of them developed this bond that says we have to have this weapon and we need to be constant and so on, as Rich said. Early in the planning, even before we got contracts, we had agreed on a set of requirements. We had a plan for our organizational and operational concept fleshed out that was understood by everybody, and we rolled it on the stage. Even 'til today it hasn't changed. Now that has to do with people setting their jaw. A few people keep singing to you and blowing in your ear saying, "Come on, we can make this thing a little bit better." But to guys like that, guys like us and guys out there at Fort Sill, we'd say, "No. We're not going to be deterred by that. We're not going to fall into that trap. We're going to develop and deploy* this *thing. Then, if there is a series of things that can make* this *thing better and more utilitarian, we'll deal with them as a planned product improvement and not screw up the development and not jeopardize getting this powerful weapon in the hands of soldiers as early as we can." So we've played absolutely stubborn. The words "go to hell" were used often, and they were understood—really. A set of requirements—the military requirements, the operational characteristics—were locked in. Then our guys had enough balls to say NO when people came in and said they were going to give us a better mousetrap.*

RDS: *Who were those people? The contractors?*

MR. SEGGEL: *Oh, it's all kinds of people. They're user people. They're headquarters people. They're guys on my staff. They're from everywhere. The guy sees an opportunity to do something better, but we said* No, No, No, No, NO, NO! *And the result is what you see! We set a master plan, which Rich put together for us, very early in the development program. That was our bible, and it went to every participant in the development program—all the government agencies and so on. Everybody had the sheet music to the symphony. And it hardly changed. It would change, maybe, as some funding variations would change, but as far as knowing what the plot was, what their role was, everybody knew. You couldn't say to me, "I didn't deliver you this on time because I didn't know I had to have it." If he said that, I'd know he was too stupid to read his own book. So we didn't have that.*

RDS: *This was almost a design-freeze type of thing?*

MR. SEGGEL: *You betcha. You betcha. . .freeze requirements! Now the contractors in the validation stage: they went off and they developed their own system to satisfy that set of requirements, and we had the shoot-off. In the end LTV [LTV Missiles and Electronics Group of Dallas, Texas] won. I would make a comment in fairness to the loser—the loser didn't lose by much. Both of them did a superb job. Now what else happened that was a key to success? We had two excellent contractors—two really professional contractor teams working the program; and after we singled down to LTV, they performed marvelously. Not all of them do that, but they did well. This program was brought in on cost. My hat's off to my contractor. My hat's off to the financial manager. It happened because it was made to happen.* There was a lot of pride involved right from the beginning. *We all started saying from the beginning, "We're going to do it better than anybody else has ever done it before. We're going to do it within cost. We're going to do it on schedule and we're going to meet the technical and operational requirements." And we kept saying that to each other every time we'd meet at quarterly meetings. We'd be looking at each other and asking, "OK, now how do we stand?" Make it, make it, make it. That went on for 64 months and we did it.*

The MLRS project's success was obviously attributable, in a large part, to the great pride, commitment, and dedication instilled by Mr. Seggel into the project team and supporting contractors, and to the continuing strong backing by management and higher headquarters.

RDS: *You attribute a lot to plain old morale?*

MR. SEGGEL: *A lotta pride—pride, pride, pride. We're better than anybody else! And we looked for ways* to do it *rather than excuses not to do it. Those weren't acceptable. We didn't play the excuses game; you had to have reasons. We didn't like to hear excuses; still don't like to hear excuses.* Now . . . All praise to higher headquarters–they didn't mess with the program. *We put the acquisition strategy up; we put up the program plan; we put the funding in front of them. They approved it (gosh, I hate to do this) but, I'm even talking about the Congress now; they supported this program; every year they put the money in the program that we needed. We were never denied what was required to do the original job set forth. All praise to 'em! They don't do that very often. But we told 'em it's going to take this many bricks, this much mortar, and this many nails, and this much timber to build this damn thing. Give it to me and we'll build it. They did, and we did!*

Even in successful projects, the unforeseen can happen. This is why the budget must have a built-in allowance for cost growth from the start. It was evident that MLRS budgeters in Washington had planned plenty of reserves, because they were ready when the unforeseen happened.

RDS: *I was going to ask you whether or not you received timely funding of the project. . .*

MR. SEGGEL: *Absolutely incredible! And it is a testimony to them and I think the Army and DOD and the Congress, when they're serious, look on it and say, "My God, look what we can do if we don't jerk the program."*

MR. PALADINO: *I think you will see three or four instances when we went back to the Congress for special funding requirements and they, through reprogramming actions, gave us the necessary funds to continue on with the program. Kind of like saying, "You're doing what you told me you would do, so I'll give you what else you need."*

RDS: *So you did have to go back for some money?*

MR. PALADINO: *Yes, we did.*

MR. SEGGEL: *We were never denied.*

RDS: *Was that due to the changes in the specification requirements?*

MR. SEGGEL: *One was a strike impact, wasn't it?*

MR. PALADINO: *One was due to a strike, one was due to a change in the test program, and one was due to the German involvement in going to the larger missile early in the program.*

MR. SEGGEL: *There were three things: Right at the beginning we were given direction by OSD [Office of Secretary of Defense] to get some European partners. We wanted some international cooperation, so in doing that we had to go and get at least Germany. We wound up getting Germany, France, and the UK [United Kingdom] and subsequently Italy. Germany, France, and [the] UK first, and then Italy several years later. But in doing that, we had to compromise on the rocket diameter and [a] little bit on range. That was early, and there was an impact to that because we had to go back and redo some things. They supported that thoroughly. That cost of those changes was retroactively to be shared. But in this case it was U.S. money up front. We had a strike at FMC which delayed us four months. That's like an act of God and there's not a thing you can do about it except pray. And then the other one was the test equipment. . . After we had gone through the validation stage, we had planned to use the HP1000, the Hewlett Packard series test equipment, a spinoff of what had been developed by Roland [the Roland Missile Project Office] and were well underway in the planning and engineering and a lot was already represented in a van that was the [Roland] item. About six months into that effort the Army made a decision that they wouldn't use that but they would use EQUATE [Electronic Quality Assurance Test Equipment]. That caused us to have to stop and go all the way back to the beginning again. . . that was a global Army decision that said the main maintenance gear for all electronics maintenance in the Army is going to be this EQUATE system. You gotta get in line. So we did.*

In our interview with Mr. Seggel, to which we will return later, it became evident that there were seven things that created an environment for success in the MLRS project: (1) *in-depth planning at the beginning*, (2) *continuity and responsiveness of funding*, (3) *strong and consistently inspiring project management*, (4) *in-depth user involvement in the requirements*, (5) *selection of an excellent contractor*, (6) *early prototyping, and* (7) *continuity of project management team personnel*. Throughout the program, high priority and commitment were evident. And in what is described by Peters and Waterman[15] as simultaneous loose–tight properties, there was a commitment to higher management to do the total job successfully without excessive, detailed involvement of higher echelons. As was the case in the Apollo moon landing program, key people in the project had stayed with it from the beginnings of exploratory development and planning through to successful completion and operation of the hardware. There was excellent technical and scientific capability in the government project management team itself, as Mr. Seggel (a civilian industrial engineer) and many of his civilian backup team had been with the project since its in-house laboratory development stages. The team knew the product inside and out; they had a great appreciation for the need for simplicity and "a user-friendliness"; and they could communicate their ideas, status, and program needs to higher headquarters to gain continued and consistent financial support. The MLRS project success was a reflection of the technical excellence, project autonomy, internal flexibility, unrestricted communications, user involvement, and sustained dedication that is possible in a project and contractor team. These attributes, coupled with the presence of solid commitment and stable backing by higher headquarters, produced an environment that not only fostered but ensured excellence.

The management and technical philosophies that fostered success in the MLRS project were the same as those which have emerged as attributes of successful projects in other defense and military programs: sustained, unwavering financial support throughout each major program phase; meticulous planning; continuity of management; willingness to listen; flexibility within the project team; and a commitment to providing the highest quality product to the user.

The Navy has prepared an excellent report that should be required reading for all government project managers and their industrial counterparts: *Best Practices: How to Avoid Surprises in the World's Most Complicated Technical Process*.[16] This book must have been used as a guideline in bringing in the aircraft carrier *Theodore Roosevelt* sixteen months ahead of schedule and $80 million under its original cost estimate. The colorfully illustrated book presents a road map for success in high-tech military weapon system development and production based on *proven* techniques and methods. Based on the results of a Defense Science Board task force on the transition from development to production, the study uses inputs from major defense industries, including General Dynamics, General Electric, Honeywell, Hughes, Northrop, Raytheon, and Texas Instruments.

The *Best Practices* study was the result of summarizing what to do from successful projects and what *not* to do from not-so-successful projects. It embodied overall dos and don'ts in six major parts of a project: design, test, production, facilities, logistics, and management. Particularly interesting are its sections on the devastating effects of erratic production and the need for stability in project management personnel. The study is applicable across the board to all major defense projects, and its keen-sighted advice is relevant to other government and large scale commercial products as well. If difficulties are encountered in a project, chances are that one or more of the "best practices" advocated in this study have been ignored. It is interesting to note that the service that has had serious financial, management, and technical problems, particularly in its shipbuilding program,[17–19] is the one that produced such an excellent study report on how to achieve success in megaprojects. Perhaps the Navy has learned much from these difficult experiences.

Now that we have learned these vital lessons, it is time to put them into practice in systems acquisition. If coupled with the broad recommendations for national systems planning, project stability, improved financial systems, and procurement law corrections presented in this book, details like those shown in *Best Practices* could send us well on our way to excellence in more programs. Synergism with government-wide management philosophies and support is required because the military services, by themselves, often have little control over some of the program drivers that have been foisted on them from the outside. Internal planning, initiatives, and controls can have only limited effectiveness if the military services are continually badgered with detailed audits, overly complex legislation, and unplanned funding fluctuations.

The Air Force's war against over-specification and drive toward program stability can be successful only if they have management backing and funding stability from above. If success is measured in terms of meeting original project cost, schedule, and performance goals, the Air Force can boast of too few successful programs of its own. During the later stages of our research, we asked the Air Force to give us some information on their most successful project or projects.[20] In early 1986 there was a lot of positive press about the excellence of the B-1B strategic bomber program. It was supposedly being delivered ahead of schedule and at a savings of one billion dollars for the first 100 aircraft. *Fortune* magazine published an article in June 1986 by Lieutenant General Bernard Randolph, then deputy chief of staff for research, development, and acquisition, that extolled the B-1B's successes, along with a statement that the C-5B transport aircraft "is coming in on schedule and at approximately half a billion dollars less than expected." *Government Executive* magazine published interviews with the Commander of the Air Force Systems Command, General Larry Skantze, and the former B-1 program manager, Lieutenant General Bill Thurman, who gave detailed rationales on why the B-1 project was such a success. All of the steps toward excellence sounded logical, plausible, effective, and transferrable to other projects. In another article, *Government Executive* magazine started a series, "Why the B-1B is Beating the Benchmarks," which was filled with detailed reasons for the program's successes.

But Frank C. Conahan, Assistant Comptroller General of the United States, gave some critical testimony about the program before a House Armed Services Subcommittee on February 25, 1987,[21] and earlier, in November 1986, an article titled "Development Problems Delay Full B-1B Operational Capability" appeared in *Aviation Week and Space Technology*.[22] The bomber was not expected to reach full operational capability for possibly two more years. The magazine cited problems such as an 80,000-pound weight growth from the B-1A version, flight stability problems, ceiling limitations, reduced range, a potential stall situation, fuel leaks, and problems with its flight electronics such as defensive countermeasures and terrain following systems. An unnamed Pentagon official told the magazine that "basically, the B-1B is a 1970-design aircraft that is being stuffed full of newer electronics equipment, and it has wire bundles all over the place." At one other point, the Pentagon official said, "The production of the B-1B bomber was a political decision, and the schedule and cost limit on the bomber has limited the changes that should have been made." Reflecting on studies fifteen years earlier of the Commission on Government Procurement, we thought, "Have we really learned anything at all from past mistakes?" Bill Thurman, the project manager, then Systems Command Vice Commander responsible for the B-1B, said, "We are finding that after 150 flight hours, the leaks are really reduced. As the aircraft come out of the Palmdale plant, we tow them around for a while to determine whether there are any leaks, and we are discovering fewer." Are we still talking about fuel leaks in aircraft that cost about $208 million each? Still, its pilots say it is a great improvement over its predecessor, the B-52. A lesson learned from the B-1B program is not to celebrate victory while the jury is still out.

Frank Conahan, in his Congressional statement in early 1987, testified that "The Air Force Test Team reported the B-1B bomber would have limited operational effectiveness at IOC (Initial Operational Capability) in areas such as navigation, terrain following, handling qualities, and defense systems. The team also reported that its weapons delivery and offensive systems would not have full capability at IOC." Mr. Conahan also reported that major logistics and support problems have developed in the B-1 program. In a two-week period in early 1987, fourteen of the twenty-five B-1B bombers then at Dyess Air Force Base were not operational for some period of time due to a lack of spare parts and inadequate repair instructions. In the same two-week period, four of the $208 million-dollar aircraft were "cannibalized" for parts to restore ten of the fourteen aircraft to an operational condition. Fault detection and isolation systems were producing false alarms 71 percent of the time. These and other technical difficulties with the B-1B bomber led Mr. Conahan to conclude that "because current problems must be corrected and significant development and tests remain to be completed, there is no assurance the baseline program can be kept at or under current estimates." Rockwell, the bomber's airframe contractor, was projecting a $500 million increase in costs above its contract target price at the time. Mr. Conahan concluded that "the high degree of concurrency between development and production was a major contributor to all the current problems." We discuss concurrency and its detrimental effects in more detail in Chapter 3.

On September 28, 1987, a B-1B bomber on a low-level bombing mission ran into a flock of birds, several of which were ingested into the engines, causing engine fires and resulting in the deaths of three crew members and complete destruction of the $208 million aircraft. There was no subsequent explanation in the press as to why a $28 billion development program had not produced bird-proof engines for a low-level bomber like the B-1B. Shortly thereafter, news articles alluding to excess profits and mismanagement of funds relative to the C-5B project began to emerge.[23] So intense were the investigations, and the smokescreen produced, that the Lockheed public relations office was not able to arrange interviews with senior or midlevel managers at Lockheed's Marietta, Georgia, plant. Despite several decades of in-depth study, analysis, reform, and innovation in the acquisition of major defense systems, negative outcomes like these have become all too familiar. As we have shown in the case of the MLRS, however, excellent weapons system acquisition projects have happened, and distinct success patterns can be drawn from these outstanding projects.

Stan Baumgartner, Cal Brown, and Pat Kelly of the Defense Systems Management College at Fort Belvoir, Virginia, conducted a detailed study of successful programs to see if lessons could be learned from their experience.[24] The MLRS was one of a dozen major military weapons systems studied to determine the ingredients that make up a successful project. This list included two Navy ships, the FFG-7 frigate and the CG-47 cruiser; the F-16 fighter aircraft and the C-141 cargo aircraft; the Polaris, Atlas, and Hellfire missiles; the CH-47 helicopter modernization; and several ground-based and airborne radar systems. Other successful projects were: Polaris I, Minuteman A, and Titan I.

The reasons cited most often for success in these projects were:

1. Good people: good office and technical staffs.
2. Good project managers, both on the government side and on the industry side.
3. A good contractor.
4. Factors related to *stability*.
   a. Realistic and stable requirements (*product* stability).
   b. *Personnel* stability
   c. *Funding* stability.

Of all the above factors, stability was found to be the most important. The report said that "product stability depends on realistic requirements (realistic for the funds available) and keeping changes to an absolute minimum. An Army project manager notes: 'Systems that have problems are those with lots of changes, especially with the user pushing for them.' Stability in funding is also essential." The report went on to say: "Vice Admiral Levering Smith of the Polaris program is admired for his frankness in advising Congressional committees on what it would cost to achieve a particular level of performance. When he was pressed to lower this figure, he explained how this would buy less performance. Over the nearly thirty years of the program, this straightforwardness has stood the test of time."

Excellent military weapons programs, as well as large nonmilitary programs, require excellence in cost estimating and a dedication to maintain a close relationship between cost, schedule, and performance. The importance of credibility and realism in cost estimating, despite pressures to become over-optimistic, will be treated in later chapters. For now, we can say that good cost estimating and control during the project are common attributes of successful programs.

Civil and Commercial Megasystems

Problems—and successes—in meeting cost, schedule, or performance goals are not restricted to the domain of large space projects or defense systems. Several studies we have reviewed, in fact, have seemed to take great comfort in the fact that defense system cost overruns, schedule slippages, or performance inadequacies have been no worse than those in civil and commercial projects of equivalent or even less complexity. (See Appendix F of the 1986 Report of the President's Blue Ribbon Commission on Defense Management.[6])

Because of the sheer size of dollar investment required, many civil projects are wholly or partially funded by the federal government. The U.S. Army Corps of Engineers alone spent $3.135 billion on civil construction projects in fiscal year 1987, in addition to $5.134 billion for military construction projects.[25] The Corps of Engineers works for other branches of the Army, other services, and other government agencies. Many Corps projects incorporate high technology and are multidisciplinary in nature. They frequently involve more than pure conventional construction.

Other civil agencies, such as the Urban Mass Transit Authority, fund up to 80 percent of the costs of public works programs such as the Washington, D.C., Metro rail transportation system. The $9.4 billion, 103-mile "Metro" has high-speed, computer-controlled high-technology trains for above- and below-ground transportation of commuters to and from the downtown Washington area and suburban locations. In our quest for successful civil systems, we launched into a study of the Metro system: its people, its policies, and its management philosophies. Although there was some cost growth in the program due to scope changes, and despite the fact that the Metro's operation is subsidized and not self-financing, anyone who has ridden in the modern, quiet, clean, and efficient trains can tell you that the system is successful in transporting commuters and visitors in comfort and with dispatch. Our first stop on the interview tour was with a key person who was involved with its construction over many years, John McAlheny[26]:

> John McAlheny, Director of the Office of Construction for the Washington, D.C., Metro (The Washington Metropolitan Area Transit Authority: WMATA) is a retired colonel of the Corps of Engineers and an experienced project manager. He has managed projects for the Corps in Iceland (B-47 refueling bases), Strategic Air Command control centers, and other large high technology projects.
>
> The planning stage of the Metro began in the 1960s. The first construction contract began in 1970. Colonel McAlheny said, first and foremost, that the reason for success was the amount of prior planning before engineering design, and then the detailed engineering design that was done before starting construction.
>
> He said that the Metro had relatively good funding stability, which was a part of the reason for the success of the project. A Congressional bill called the Stark-Harris bill provided the federal portion of the funding. The 89-mile portion of the Metro is to be completed by 1994. (They now have 69.59 miles in revenue service). The ultimate system will have 103 miles in service. The Metro uses a "cost-loaded CPM" (critical path method) procedure for planning each section of the line. Each schedule element is loaded with cost in order to time-phase costing. Their contracts are broken down into three separate bid items. Originally they had about 100 separate bid items, but they have consolidated these so that now they have only three separate bid items: mobilization (the effort required to set up for construction), basic construction, and safety awareness.
>
> As to lessons learned, *prior planning* is the key: setting up realistic design and construction schedules, doing very detailed design, and giving the contractors adequate design detail to complete the work. The Metro's construction is organized into (1) the brick and mortar segment and (2) all other things such as communications, automated train control, power, and equipment.
>
> Another reason for the Metro's success is that they didn't hesitate to borrow from foreign technology. They reviewed all of the world's urban mass transit systems prior to starting the concept formulation of the Metro, using the very best of every one, and rejecting the worst. They are using a tunnelling technique, for example, which was developed by the Austrians.
>
> They are continually studying other urban mass transit systems that have been built or are underway, such as the BART (Bay Area Rapid Transit) system in San Francisco, the MARTA system in Atlanta, and a much smaller but similar system in Baltimore.

Although the GAO prepared a report in August 1983 that was generally critical of the Metro's procurement of rail cars,[27] the WMATA (Washington

Metropolitan Area Transit Authority) and Department of Transportation responses to the report and corrective actions taken appear to have corrected problems in this area. Some slippages in delivery schedules of individual lots of cars occurred, but the final product seems to have acceptable quality and safety!

When asked about long-range planning beyond the 103-mile system, John McAlheny said there are definitely spurs that are needed (for example, out to the Dulles Airport), but there has been very little planning done beyond the 103-mile system.

Every two to four weeks they have a scheduling meeting to review the critical path schedules posted on the wall in John's office. Metro experts plus domestic and foreign advisors assess and comment on the progress. Meticulous planning is evident. One point mentioned, again and again, was don't hesitate to go outside the country to get the best experts if they are needed.

We mentioned that the system was very user-friendly, and he said, "Yes, that was because they visited all the major metro systems in the world and selected only the best criteria." Little details were important, such as the station arches being set back from the platform so that people can't reach them to add graffiti. Eating, drinking, and smoking are not allowed; thus, rodents, cockroaches, odors, and litter are kept to a minimum. We noticed that there are no exposed light bulbs in the Metro. Indirect lighting is used so that no one can steal the light bulbs or shoot them out.

Another reason for their success, according to John, is that they have had two long-term consultants that stayed with the project many years. The general engineering consultant company has been with them since the 1960s. Metro also hired a general construction consultant, Bechtel, to give continuity to the program through construction. Bechtel was with the project from 1970 all the way up through 1985. Finally, in 1985, the general construction supervision was brought in-house.

We asked Colonel McAlheny about the management of the Metro. The Board of Directors of the Metro Authority is made up of officials from Maryland, Virginia, and the District of Columbia—some are appointed; some are elected. The general manager of the Metro, Carmen Turner, is a superb administrator. She has five deputies: one for business administration, one for business services, one for finance, one for rail services, and one for design, construction, and facility maintenance. There are 323 people in the construction division.

When we asked Colonel McAlheny about testing of the system, he said they go through a one-year complete detailed test of the system once it's in place. Then they have a seven-month operational test prior to transporting passengers. Colonel McAlheny said that when he came on board in 1981, there was an operational date set for the system, and it has been met. Over a period of about six years, the scheduled dates have stayed firm, and they have been able to meet their schedule. They have a very unique and strong safety program, and they are self-insured.

John McAlheny also said that one of the significant reasons for the Metro program's success was that they had so many people that came over from the Army Corps of Engineers. The key people have an engineering background; salaries of the Metro agency are a little higher than normal, which attracts good people. They have good technical and business people to run the program.

It was interesting to note in this interview, as well as in a later interview with a project manager, Dave Schilling, how important it was to use the best technology and advice available, regardless of its source. Of all of the three excellent projects we studied (Apollo, MLRS, and Metro), none hesitated to use foreign technology, ideas, or even foreign personnel. Although all three of these projects did encounter the normal difficulties in development and procurement, agency managers, acquiring help wherever they could get it, were able to overcome these problems to produce programs that were judged to be highly successful when compared to the average federal major system. NASA not only imported technology along with the German-born scientists, but is still involved in heavy cooperation with the Europeans and others in its Spacelab and space station projects. Mr. Seggel has Europeans working on his team as an integral part of his MLRS project. And John McAlheny uses the Austrians as key members of his government/industry crew.

We noted an interesting atmosphere at the Metro—an enthusiasm and dedication stemming from, we believe, a desire and willingness of management to foster innovative actions on the part of the construction managers, resource managers, and operational personnel. Project managers for various aspects of the job were given both the responsibility and authority to make on-the-spot decisions to save money or to improve quality, even though it may have meant some internal redesign of their phase of the work. The engineers, accountants, and administrators knew that they had the backing of management to innovate and improvise. They were managed by being given the overall goals and philosophies of cost consciousness, service to the public, quality, and safety, and not a rigid set of rules, regulations, approvals, and controls.

All of the exemplary projects had plenty of technical people on their project management teams. Engineering management studies consistently point out that the key personnel in large, complex projects must have in-depth general backgrounds if not intimate working knowledge of the product or system being developed.[28,29] Each of the key people we interviewed said that a major contributor to success was having high-quality technical people in the project office and *on the job site*. Diligent, detailed planning, willingness to say *no* to disruptive, externally imposed changes, and program flexibility to work *within* the programs to bring about cost avoidances and cost savings were also common attributes of these three widely divergent but extraordinarily successful programs.

Another notably successful large-scale project that is strictly a commercial venture but that deserves some attention from its public sector counterparts is the development and construction of the $1 billion EPCOT Center near Orlando, Florida—a high-tech, engineered mini-universe where visitors can thoroughly experience the past, present, and future for the price of a three-day ticket. The computerized entertainment center of 260 acres and 19 complexes of exhibits and buildings (pavilions) cost less than a single

Trident submarine. There are lessons to be learned in its planning, construction, and operation applicable to high-tech military, space, and civil projects undertaken by the U.S. government. A host of scientific, engineering, and technical disciplines had to be mustered to develop this model information society of the future. Many of the same techniques used to plan and design our advanced weapons and their bases were used to develop EPCOT. For example, modern computerized scheduling, estimating, and industrial engineering techniques were used for planning and construction. The details of how the Walt Disney Productions staff "imagineered" a high-tech, innovative, highly successful complex like EPCOT is a story in and of itself. Suffice it to say now, however, that technology transfer may be coming from the commercial sector back to the public sector just as it has been traveling the *other* way for many years.

How Foreign Governments Buy Megasystems

How do countries such as France, the United Kingdom, West Germany, Israel, and the Soviet Union acquire their major systems? And can we learn anything from these countries in terms of developing the anatomy of a successful project? The answers to these questions are partially revealed in a fascinating 1986 GAO report titled *Weapons Acquisition: Processes of Selected Foreign Governments*.[30]

It is interesting to note that in this report all four of the free world countries evaluated in the studies (which have an average of 7 percent of the military budget of the United States) prepare and update long-term plans ranging from ten to fifteen years and approve firm detailed budgets for four or five years for their major defense acquisitions. The United States Defense Department has a five-year defense plan and only a one- to two-year detailed and approved budget.

Do countries with fewer resources plan their use more effectively? Have they found that a four-to five-year budget is required to achieve project funding stability? The answer to both questions is most probably yes! Acquisition organizations differ widely among the countries studied, heightening our conviction that it is not the organizational arrangement that solves the problems (not that the other countries do not have any problems). But, apparently, a wide variety of organizations will work. The section of the report on system acquisition in the Soviet Union did not say whether Russia has a multiyear plan, nor did it specify its length (although from what we have heard before about the Russians, it is probably of twenty-year or maybe even fifty-year duration). However, a few important excerpts from GAO's comments in the Soviet portion of the report are given below.

> *The Soviet design philosophy of simplicity, standardization, and inheritance minimizes uncertainty in development and costs, and limits problems in production.*

The Soviet acquisition process is a sequence of disciplined, risk-avoiding steps.

Tight deadlines are imposed.

Approval of each step completed is marked by the joint signatures of system designers, government monitors, and military customers.

Design agreements are legally binding.

Accountability is clear.

No production in Soviet plants begins until development and testing is finished.

Concurrency (overlapping among development, testing, and production) is unheard of.

Cost overruns are not tolerated by the hierarchy.

Officers specially trained in system acquisition are present in research institutes, design bureaus, and production plants to make sure that their subject systems meet military requirements.

The military officer on the site is directly responsible for quality. He can shut down the production line and refuse the product.

Production does not begin until the prototype is proven.

Prototypes undergo extensive laboratory and factory trials conducted by the design team, factory management, and the military customer to check adherence to the tactical technical instruction.

The designer has a large say in production and is held chiefly responsible for quality and schedule.

Uncomplicated designs maximize standardization opportunities, enhance reliability, cost less, and reduce maintenance, training, and logistics.

An interesting point is that its (the MiG-21's) new engines were retrofitted into earlier aircraft, keeping them in inventory.

Concepts now being discussed in the United States with increasing frequency—prototyping, reducing or eliminating concurrency, freezing or baselining designs, upgrading existing rather than developing new weapons, improving the project manager's authority, and tightening budgetary controls—appear parallel to some of the techniques the Soviets are finding beneficial. Many of these same approaches were found in the other countries studied and are parallel to our own findings and recommendations. With this type of exposure of how others operate, the United States should take the initiative to select those ideas that work—and reject those that don't—in its major weapons systems.

What about large-scale civil and commercial projects in other countries? Dr. Peter W. G. Morris, a research fellow at Templeton College in Oxford, England, prepared an excellent research paper titled "Lessons in Managing Major Projects Successfully in a European Context."[31] Morris looked at projects as widely varied as the proposed English Channel Tunnel, the Concorde supersonic aircraft, the Fulmar North Sea oil field, and the European Giotto spacecraft to determine the elements that enhance large scale project success in Europe. One of the most succinct statements of advice in his report was stated in a very British way in only two words: "Avoid rushing!" His deductions are consistent with the key discoveries of many other studies: avoid concurrency; develop good planning; phase projects; design comprehensively, freeze design, ensure continuing sponsorship, beware of the cheapest bid, seek competent personnel, obtain strong leadership, establish firm commitments, and many more. His final conclusion, stated succinctly in one sentence, was

> *If there is one overall sense of what emerges as important, it is commitment, from the highest levels down, married to the utmost care in planning, financing, design, testing, and contracting; with leadership and teamwork from the highest levels to the work force; implemented with care, persistence, and with clearly marked milestone stages.*

If there is one sentence to sum up the anatomy of a successful project, we would use the above.

A Quest for Excellence

In July 1985 President Ronald Reagan established the President's Blue Ribbon Commission on Defense Management to study defense management policies and procedures which produced the report titled A Quest for Excellence.[6] The charter and activities of this group were similar to those of the earlier 1969–1972 Commission on Government Procurement[32] except that: (1) the study was limited to the Department of Defense, and (2) the study was made of the *overall management* of the DOD rather than just procurement. The Commission found "an urgent need for reforms in defense management." Its paramount purpose was to "*identify and develop solutions to structural problems in the Department of Defense.*" The Commission found a requirement to "*cultivate resilient centers of management excellence*" dedicated to advancing DOD's overall goals and objectives.

David Packard, Commission chairman, in his foreword to the Commission's final report, reached the following inspiring and profound conclusions:

> *Excellence can flourish, I believe, only where individuals identify with a team, take personal pride in their work, concentrate their unique efforts, develop specialized know-how, and, above all, constantly explore new and better ways to get their jobs done. Freedom and incentives of just this sort, President Reagan has observed, "unleash the drive and entrepreneurial genius that are the core of human progress."*

Mr. Packard went on to say that the *centers of excellence* approach can tangibly improve productivity and quality and achieve revolutionary progress throughout defense management. Mr. Packard also concluded that excellence in defense management cannot be invoked by legislation or directive. Excellence requires the opposite—responsibility and authority placed firmly in the hands of those at the working level who have knowledge and enthusiasm for the tasks at hand. Mr. Packard said that excellence in defense management will not come from legislative efforts to control and arrange the minutest aspects of DOD's operations and that "Congress must resist its inveterate tendency to legislate management practices and organizational details."

We observe that project management teams for megaprojects in the Department of Defense as well as in other arms of government can form the *centers of excellence* referred to by Mr. Packard. Since the project team is identifiable with a specific project or task, it can be given the overall job, along with financial and management support, and can be relied on to produce results without excessive rules, regulations, controls, approvals, and micromanagement. To do this, the inspiration and commitment must come from the top down. Senior managers must be willing to establish superordinate goals and to relinquish detailed control to achieve excellent results.

Centers of excellence for megaprojects have already been proven effective in the form of project teams like the Apollo propulsion development team, Mr. Seggel's MLRS weapons project office, the Washington Metro construction team, and many others. Our goal should be to reach a point where every project team for every megasystem is a center of excellence.

In major systems acquisition for the U.S. Government, the new philosophy means that even the Congress must concentrate on the larger, often neglected issues of long-term national project planning and goals rather than dictating, through legislation, the detailed means of achieving these goals. Strategic plans are needed that focus upon the issues.

The Packard Commission report states: "The entire undertaking of our national defense requires more and better long range planning. This will require a concerted action by our professional military, the civilian leadership of the Department of Defense, the President, and the Congress." This long-range planning must be done with a profound knowledge of what really *works* in the acquisition of major systems. As the Packard Commission also points out, the quest for excellence will be successful only with a new management philosophy. Instead of concentrating on the things that are being done wrong and trying to fix them with more laws, regulations, and inspectors, we should concentrate on those things that are done right and use them as models.

One such group of models are the excellent corporations in the United States that have carried out mammoth development, construction, production, and information projects using private capital. How these companies successfully carried out enormous projects is well documented in a group of "excellence" books (*In Search of Excellence, A Passion for Excellence, Managing for Excellence, Creating Excellence,* and others) that emerged in the mid-1980s. The Defense Science Board and an Acquisition Task Force formed by the Packard Commission studied several DOD programs that were developed under special "streamlined" procedures—the Polaris missile, the Minuteman missile, the air-launched cruise missile (ALCM), and several highly classified projects. The Commission found that, in these programs, the Department of Defense equalled the accelerated schedules achieved by excellent commercial programs. In search of a pattern that could be applied to defense programs, the Packard Commission found a model to emulate. They discovered the following underlying features that typified the most highly successful commercial programs:

- *Clear Command Channels.* The project manager must have singular responsibility and accountability for his or her program, and a short, unambiguous chain of command to the Chief Executive Officer (CEO). Groups or individuals wishing to influence program actions must provide inputs to the responsible project manager, who may accept or reject their proposals. Major unresolved issues are referred to the CEO, who has the final say-so in resolving conflicts. To emulate industry in this area, the government will have to minimize layering by holding the project manager directly responsible to line chief executives. [Ironically, the Packard Commission also recommended (as was subsequently adopted) a new Undersecretary for Acquisition in the Department of Defense who is *interspersed* between the project manager and line chief executives.]
- *Program Stability.* At the outset of the program, the project manager enters into a fundamental agreement or "contract" with the CEO on *specifics* of performance, schedule, and cost, and the CEO provides strong management and funding support throughout the life of a program. This gives the project manager a maximum incentive to make realistic estimates and maximum support in achieving them. In turn, the CEO does not authorize full-scale development until the board of directors is (1) solidly behind it (which means that extensive and meticulous planning, scheduling, and estimating have been done), (2) prepared to fund the program fully (through its full current phase), and (3) willing to let the CEO run the program within the agreed-to funding (the Board of Directors does not get involved in the day-to-day, week-to-week, or month-to-month management of the project). Incentives and peer reviews are needed in both government and industry for project managers to criticize aspects of their own programs rather than to merely reward project managers for being advocates of their own programs.
- *Simple Reporting Requirements.* The project manager reports only to the

CEO. Typically, this is done on a "management-by-exception" basis where progress is focused principally on deviations from the plan.

- *Small, High-Quality Staffs.* Personnel are hand-selected by the project manager and are highly capable in both business and technical disciplines. Program staffs spend their time managing the program, not selling it or defending it.
- *Communications with Users.* The project manager establishes a dialogue with the customer, or user, at the conception of the program when initial trade-offs are made, and maintains that communication throughout the program. When development problems arise, performance trade-offs are made with the user's concurrence, but not in-line approval, in order to protect cost and schedule commitments. As a result, the project manager is motivated to seek out and address problems rather than hide them.
- *Prototyping and Testing.* End items, systems, or critical subsystems are (1) simulated, designed, and constructed in laboratory, scale model, and/or prototype form; (2) tested under credible operational conditions; (3) modified to correct deficiencies found in testing; and (4) subjected to final qualification testing before final design approval or authorization for production. Prototypes should be used in early advanced development as well as later in full-scale development. In many cases, the project manager establishes a "red team" (a "devil's advocate") or peer review involving persons within or external to the program office to seek out pitfalls in the prototype design—particularly those that might arise from operational problems, or from an unexpected response by a competitor. Prototyping, early operational testing, and red-teaming are used for the timely identification and correction of problems unforeseen at the program's start. Although competitive prototyping as an acquisition strategy was not formally recorded as a technique used in the commercial marketplace, it has been beneficial in some government acquisitions as a means of maintaining "competition" until later program phases. Prototypes for later phases must closely resemble production items and must be tested prior to the start of production to be of maximum benefit. Risks must be identified and dealt with before significant funds are committed to a program.

Does the Packard Commission Report Describe the Anatomy of a Successful Project?

In reviewing the Packard Commission Report, we found that the problems and solutions it presented were not applicable just to those in defense management; they spread across the entire spectrum of major systems acquisition for the federal government. To go even further, we are talking about major systems funded in part by the federal government, although many of the same problems and solutions are applicable to state and local governments as well as to the commercial or private sector. The Commission found an "urgent need for reforms in defense management" and stated early that this would be accomplished by "a new management philosophy rather than major institutional change." David Packard, in his foreword to the report, twice mentioned proposed centers of excellence and said that excellence in defense management "cannot be achieved by the numerous management layers, large staffs, and countless regulations in place today." Yet the report itself did not pursue the centers of excellence approach and actually recommended a new level in the hierarchy to manage systems acquisition.

We believe there must have been a breakdown in communication between the chairman and the supporting commission staff and organizations because there was no strong and overlying directive to simplify organizations; to use existing people, facilities, and resources; or to change management philosophies rather than organizations. The report's recommendations were intended to help establish strong centralized policies that were to be both sound in themselves and rigidly adhered to throughout the Department of Defense. Cultivation of resilient centers of management excellence, a central recommendation of Mr. Packard, was not described, planned, or established in the report and appears not to have been widely implemented over the many months since the report was issued. Mr. Packard said that ways must be found to restore a sense of shared purpose and mutual confidence among the Congress, DOD, and industry, but the report did not describe how to restore this confidence.

The Packard Commission recognized that the entire undertaking of our national defense requires more and better long-range planning, but it continued to advocate only a five-year plan, which is not sufficient for current major projects, as will be shown in the next chapter. The Commission concluded that new procedures are required to help the administration and Congress do the necessary long-range planning and meaningfully assess what military forces are needed to meet national security objectives; yet it did not describe or recommend workable procedures to do so. The report said that impressive savings would come from eliminating the hidden costs that funding instability imposes; yet it made no new recommendations as to how to eliminate funding instability.

The Packard Commission study revealed—and the collective experience of the Commission members fully confirmed—that there are certain common characteristics of successful commercial and government projects: (1) short, unambiguous lines of communication among all levels of management; (2) small staffs of highly competent professional personnel; (3) an emphasis on innovation and productivity; (4) smart buying practices; and, most importantly, (5) a stable environment of planning and funding. In our view, the Packard Commission should have developed a recommended plan of action and a scenario as to how each of these objectives could be accomplished over the coming years.

In the area of procurement, the Packard Commission recommended a single consistent and greatly simplified procurement statute. This is the same recommendation that was made by the Commission on Government Procurement in 1971 (sixteen years earlier) and again by the Office of Federal Procurement Policy in 1982.[33] In this case our question is not how the Packard Commission or the DOD would implement it, but why this simplified procurement statute hasn't already been developed as a result of previous recommendations to the Congress and to the President.

In the area of testing, the Packard Commission report said that the project manager should agree to a baseline for all phases of his program. For the acquisition executives, however, the agreement should extend only to full scale development and low-rate production. Before a program could enter high-rate production, it would be subjected to developmental and operational testing. This third phase is far too late for development and operational-style testing. Testing must start in the earliest program phase and be an integral part of all program phases until the system is fully ready and qualified for use.

There is a lot about contractor "self-governance" in the Packard Commission Report. Granted, giant steps have been taken in improvement of government contractor ethics. But to depend more heavily on contractor internal controls and codes of ethics while removing punitive measures such as suspension and debarment for infraction of federal procurement statutes seems to be an action of less than the best business wisdom. To many in the defense industry, debarment (prevention from participating in government contracts)—whether it be temporary or permanent—may seem like unjust punishment. But, in the commercial world, is it considered unjust punishment when a consumer refuses to buy from a company which has been routinely dishonest, or which consistently provides overpriced, shoddy goods or services to its customers? No. It is through this unregulated competitive environment that the superior performers emerge. If we are truly trying to emulate the efficiencies of the private sector, we must also absorb the concurrent requirement for long-term customer satisfaction. The report said, further, that contractors have a legal and moral obligation to disclose to government authorities misconduct discovered as a result of self-review. But is it realistic to assume that they can be motivated to do so? For example, the Commission was concerned that overzealous use of investigative subpoenas by Defense Department agencies may result in less rigorous internal corporate auditing. The report said that ways must be found to attract highly qualified personnel from industry, yet it recommended that ways be found to close or slow down the "revolving door" (swapping of industry and government people).

In the internal deliberations of the Packard Commission, there must have been much disagreement as to whether reorganization in DOD should be avoided at all costs or actively sought after as a means of implementing positive reform: "While otherwise convinced that the Secretary should be left free to organize his Office as he sees fit, the Commission concludes that the demands of the acquisition system have become so weighty as to require organizational change within that office."

The Commission report said that a better job of determining requirements and estimating costs is needed at the outset of weapons development to reduce overruns. Recent studies have shown that 15–25% of the overrun is a result of cost estimating errors. At least part of the overruns in DOD actually result from faulty estimates at the beginning of the programs. Recent GAO reports and their responses from DOD indicate a need for uniform and improved cost-estimating systems both in DOD and other agencies and at DOD contractors.[34–37] One of the sixteen appendixes included in the Packard Commission Report addressed cost estimates. It explained that cost estimates in military programs are "no worse" than those in commercial programs. Recommendations were needed that would provide specific means of improving cost estimating by the DOD and its contractors.

Despite some inconsistencies in the Packard Commission Report, many of the basic concepts advocated are potentially beneficial and agree in substance with our own conclusions and with the principal recommendations of other studies. These include increased involvement of the President in national security planning, reliance on inherent market forces rather than government intervention to increase competition, improving financial accountability, giving the project manager full authority to carry out his or her program, and enhancing program stability through authorization and appropriation of major programs not annually but at key milestones. Studies of the anatomy of successful programs have shown that these factors, although they have not all been present in all successful programs, are major contributors to program success.

Some Not-So-Successful Civil Projects

While we learned what *to do* from the successful projects, have we also learned what *not to do* from the not-so-successful megaprojects? And do the findings of the two viewpoints coincide?

As a case in point, let us look at a recent General Accounting Office report titled *Cost Escalation in Three Major Department of Transportation Projects*.[38] The report states:

> *In recent years, a number of Department of Defense and civilian agency projects have exceeded their original cost estimates. The Congress used these estimates to make funding decisions. When project costs are higher than originally anticipated, federal*

agencies must either obtain more funds, buy less, or stretch out the completion of the project, which often increases total project costs.

Enormous cost growths had been experienced in the Coast Guard's Short Range Recovery Helicopter, the Federal Aviation Administration's Airport Surveillance Radar, and the Urban Mass Transportation Administration's Buffalo Light Rail Rapid Transit System. The cost growth *above original rough estimates* for these three projects was 166 percent, 263 percent, and 59 percent respectively. (If the original estimate is *included* in the percentages, they would have to be expressed as 266 percent, 363 percent, and 159 percent of their *original* estimates.) Together, these three projects began with a cost estimate that totaled about $663 million. By the time the projects were completed, they collectively totaled more than $1.5 billion.

The GAO set out to discover why the cost growth occurred. Reasons for the cost growth included unanticipated inflation, engineering changes (a change in the *definition* of the item being acquired), quantity changes (changes in the *number* of items being acquired), start-up costs, inadvertent omission of other required items from the original estimate, schedule delays, socioeconomic law enactments, and "unknown causes." (The GAO listed $11.4 million of cost growth due to "unknown causes" in the Buffalo Light Rail Rapid Transit System). In the overall analysis, the cost increases were due to inadequate overall planning, changes in scope of the projects, inadequate definition, insufficient allowance for inflationary costs, and just plain underestimating. In GAO's view, preliminary proper planning, definition, and estimating, along with close tracking of inflationary indices *during* the projects, would have resulted in commencement of only those projects that were affordable. Better management and control of those that were approved would have been possible if firm budgets with adequate allowances were established at the beginning.

Project managers, procurement people, and even Congressional staffers have told us that a continuing uncertainty as to whether their program is going to be in the budget next year is, in itself, a cost driving factor. Uncertain and fluctuating funding has been blamed for cost growth of from 30 to 50 percent in selected projects because the uncertainty of funding each year is, in itself, an unsettling factor that accounts for uneconomical starts and stops, speed-ups, and slow-downs in a program.

One or more of several conditions has created many not-so-successful government and government-sponsored megaprojects:

- Inadequate long-range, in-depth, national planning that clearly establishes timing, priorities, and resources for major systems; and inadequate in-depth definition, analysis, and planning of major systems themselves.
- Lack of a firm and lasting financial commitment to fund each project through stepwise, incrementally approved phases; and lack of a corre-

sponding visible, methodical, timely, and meticulous government-wide accounting, estimating, and financial reporting system.

- Inadequate motivation, authority, continuity, training, experience, skill mix, and/or selectivity of project managers and project management teams.
- Counterproductive and confusing procurement legislation.
- Failure to subject complete systems to rigorous operational test and evaluation before the start of production.

These conditions, along with synthesized approaches to their correction, are discussed in detail in the following five chapters:

Chapter 3: A Dynamic National Program Plan
Chapter 4: Establishing Financial Stability
Chapter 5: The Project Manager
Chapter 6: Private Industry and the Procurement Process
Chapter 7: The Role of Congress

We believe that observing the anatomy of successful projects and reviewing what others have discovered about large-scale, high-dollar-value, complex, high-technology programs can result in formulation of a workable methodology, management philosophy, and framework for action, the adoption of which will ensure success in government big-ticket acquisitions. These suggestions are included in the remaining chapters and summarized in Chapter 8.

Endnotes

1. Frederick I. Ordway and Mitchell Sharpe: *The Rocket Team,* MIT Press, Cambridge, MA, 1982.
2. Interview with Eberhard Rees, Director (retired): Marshall Space Flight Center, NASA, Huntsville, AL, July 24, 1986.
3. Presidential Commission on Space Shuttle: *Report of the Presidential Commission on Space Shuttle Challenger Accident,* Presidential Commission on Space Shuttle, Washington, DC, June 1986.
4. Institute of Cost Analysis: *Proceedings of the First Annual ICA Symposium,* Institute of Cost Analysis, Washington, DC, December 1982.
5. Michael Rich & Edmund Dews with E. L. Batten, Jr.: *Improving the Military Acquisition Process,* Rand Corporation, Santa Monica, CA, February 1986, R-3373-AF/RC.
6. Packard Commission: *A Quest for Excellence: Final Report of Packard Commission, 2 Vols.,* Packard Commission, Washington, DC, June 1986.
7. General Accounting Office: *Recommendations on Department of Defense Operations,* General Accounting Office, Washington, DC, February 1985, OIRM-85-2.

8. Office of Management and Budget: *Executive Order 12352 on Federal Procurement Reforms: A Reform '88 Program*, Office of Management and Budget, Washington, DC, February 1984, M-84-7.
9. Lester Fettig: "Defense Management Reform: The Carlucci Iniatives a Year Later," *Military Electronics/Countermeasures*, PASHA Publications, Arlington, VA, May 1982.
10. Commission on Government Procurement: *Major Systems Acquisition—Final Report*, Commission on Government Procurement, Washington, DC, January 1972.
11. U.S. Congress: *Hearings on Federal Spending Practices, Efficiency, and Open Government*, U.S. Congress, Washington, DC, 1975.
12. General Accounting Office: *The Army's Multiple Launch Rocket System is Progressing Well and Merits Support*, General Accounting Office, Washington, DC, February 1982, MASAD-82-13.
13. Jerry Stilkind: "MLRS Stands Out as a Procurement Success," *Army Times*, Army Times Publishing Co, Springfield, VA., November 25, 1985.
14. Interview with Larry Seggel, Project Manager: Multiple Launch Rocket System (MLRS), Army Missile Command (MICOM), Huntsville, AL, September 5, 1986.
15. Thomas J. Peters and Robert H. Waterman, Jr.: *In Search of Excellence*, Harper & Row, New York, 1982.
16. Department of the Navy: *Best Practices: How to Avoid Surprises in the World's Most Complicated Technical Process*, Department of the Navy, Washington, DC, March 1986, NAVSO P-6071.
17. Jacob Goodwin: *Brotherhood of Arms*, Times Books, New York, 1985.
18. Time, Inc.: "The Navy Under Attack," *Time*, Time, Inc., New York, May 8, 1978, Vol. III, No. 19.
19. Time, Inc.: "The Winds of Reform," *Time*, Time, Inc., New York, March 7, 1983.
20. Department of the Air Force: *Air Force Acquisition Statement FY87*, Department of the Air Force, Arlington, VA, 1986.
21. Frank C. Conahan: *The B-1B Aircraft Program* (testimony to Armed Services Committee), General Accounting Office, Washington, DC, February 1987, T-NSIAD-87-4A.
22. Aviation Week: "Development Problems Delay Full B-1B Operational Capability," *Aviation Week and Space Technology*, McGraw-Hill, New York, November 1986.
23. Norman Black: "$2.2 Billion Contract Nears OK," *Huntsville Times*, Huntsville, AL, September 26, 1986.
24. Defense Systems Management College: *Program Manager*, Defense Systems Management College, Fort Belvoir, VA, January-February 1984, Vol. XIII, No. 1.
25. U.S. Army Corps of Engineers: *Program Projections*, Headquarters, U.S. Army Corps of Engineers, Washington, DC, October 1986.
26. Inteview with John McAlheny, Director of Construction Office: Washington Metropolitan Area Transit (Metro), Washington, DC, August 26, 1986.

27. General Accounting Office: *Metro Needs to Better Manage Its Railcar Procurement*, General Accounting Office, Washington, DC, August 1983, NSIAD-83-26.
28. Gary Hart and William S. Lind: *America Can Win*, Adler & Adler, Bethesda, MD, 1986.
29. Commission on Government Procurement: *Special Report on Acquisition of Major Systems—Draft*, Commission on Government Procurement, Washington, DC, December 1972.
30. General Accounting Office: *Weapons Acquisition: Processes of Selected Foreign Governments*, General Accounting Office, Washington DC, February 1986, NSIAD-86-51FS.
31. Peter W. G. Morris: "Lessons in Managing Major Projects Successfully in a European Market," *Colloquium on Research Priorities for Large Scale Programs 1985-1*, Large Scale Programs Institute, Austin, TX, March 1985.
32. 91st Congress: *Commission on Government Procurement Public Law 91-129 83 Stat 269*, 91st Congress, Washington, DC, November 26, 1969, PL 91-129.
33. Office of Federal Procurement Policy: *Proposal for Uniform Federal Procurement System*, Office of Federal Procurement Policy, Washington, DC, February 1982.
34. Office of Management and Budget: *DOD Needs to Provide More Credible Weapon Systems Cost Estimates to Congress*, Office of Management and Budget, Washington, DC, May 1984, NSIAD-84-70.
35. General Accounting Office: *Selected Civilian Agencies' Cost Estimating Processes for Large Projects*, General Accounting Office, Washington, DC, September 1986, GGD-86-137BR.
36. General Accounting Office: *Contract Pricing/Defense Contractor Cost Estimating Systems*, General Accounting Office, Washington, DC, June 1987, NSIAD-87-140.
37. Robert B. Costello: *DOD Response to GAO/NSIAD-87-140*, Department of Defense, Washington, DC, September 1987, P/CPF.
38. General Accounting Office: *Budget Issues: Cost Escalation in Three Major Department of Transportation Projects*, General Accounting Office, Washington, DC, July 1986. AFMD-86-31.

CHAPTER THREE

A Dynamic National Program Plan

Often when an individual or organization attempts to attack a problem of great magnitude, it is necessary to look *beyond* that individual or organization for root causes and solutions as well as *within* the entity that is performing the work. Much of the past research on acquisition, procurement, contracting, and project management techniques has concentrated on what the contracting officer or the project manager can do to correct the problem of large unanticipated cost increases in major projects. We and others[1,2] have shown that successful programs can and do exist. Excellence can exist within the current framework of checks and balances in our government. The problem is, as one official in GAO told us, program success in terms of meeting or exceeding performance goals within the bounds of estimated time and cost is rare. Excessive costs and unsatisfactory performance have continued to be major problems in the acquisition of some of our major systems and projects. Therefore, we must look both into and outside of the organizations that are used to acquire these systems to find workable solutions.

This chapter and the following one address needs that are becoming evident: needs that reach far beyond the day-to-day workings of a project office and into some fundamental actions that must be taken by the Congress and the President to ensure that *all* major systems can eventually be procured and operated with efficiency, economy, and effectiveness. In this context, the two areas for potential improvement that we will address first are the need for a *dynamic national program plan* for major systems and the need for an *industrial-type, budget-sensitive, management accounting system*. These two needs, as will become evident when we discuss problems, potentials, and possibilities, must be an integral part of any long-term initiative for the improvement of major systems and projects. Without them, we will most likely retain only pockets of excellence where unique individuals and unique situations have brought about the most desirable outcomes. Although it is the highest priority need and must be accomplished first chronologically, we will address the government-wide accounting system in the *next* chapter. Because it may be the most controversial and seemingly difficult task, the establishment of national goals, plans, and multiyear budgets will be discussed in this chapter.

The public and the Congress are becoming more fully informed of the detrimental effect of large federal deficits and a growing national debt. Management of the government within available resources is already becoming somewhat of a national goal in itself, and it won't be long before it becomes a reality. In the meantime, federal project managers are crying out for program stability. Government studies have shown that program instability is the most significant cause of the outrageous cost of major systems.[3,4] To gain program stability in an atmosphere of fiscal responsibility, there will be no alternative but to institute a national goals structure and to develop and maintain a dynamic national program plan. The need for an integrated, long-range plan for major systems became clearly evident in early 1986 when the space shuttle Challenger accident in late January was followed by the Titan 34D loss in mid-April of the same year. The nation suddenly discovered that it had no reliable space launch vehicle. There was no one in the Congress or in the administration who was institutionally or jurisdictionally placed to examine the combined impacts of Titan and Challenger so that national space policy and major system development could be altered accordingly. To provide stability and to coordinate major system and project goals, evidence points toward the need for a twenty-year plan that contains the track record of the past five years plus a plan for the next fifteen years of major systems development and production, and that this plan should be accompanied by a framework of multiyear budgets for major systems.

Long Programs Require Long-Term Planning

The length of time it takes to develop, build, and make operational our major systems is ten to fifteen years.[5,6] A five-year planning process can only cover a third to a half of the acquisition life cycle of a major system, and a one-year budget is insufficient to cover even a major program phase. A twenty-year plan will envelop many confluent programs and will adequately encompass the major phases of even the largest projects. As a group, long-term projects that have enjoyed sustained support from the Congress, the President, and the public have good track records of success, while those with sporadic support have not been cost-effective.

Ten- to fifteen-year program durations have been with us for a long time. Dick Reynolds of the Defense Advanced Research Projects Agency told us that the time span from idea germination to operational readiness for an innovation is twelve to fifteen years. He cited the example given in Chapter 1 of the cannon lock invented by Lt. Dahlgren in 1835[7]: It took ten years for the device to be designed into an operational weapon and five more years before the weapon could be used widely in the field. Dick also provided a more recent illustration, the gallium arsenide transistor which was invented in 1964 but not widely used in systems until 1978—a span of fourteen years.

Present-day high-tech weapons systems from aircraft to submarines typically take seven to fifteen years from concept formulation to operational readiness. NASA's space station development will span a six- to eight-year period after being studied in detail for two to four years—a total time span of ten years. The Air Force's advanced tactical fighter aircraft, in concept studies since 1984, was selected for prototyping in 1986. The prototype will fly in 1989/1990; the aircraft will be put into production in 1993, and it will obtain operational status by 1996—a total of twelve years since the initial concept studies. Big civil construction projects such as the trans-Alaska gas pipeline; the Department of Energy's 64-mile diameter, $5 billion superconducting collider facility; a new nuclear waste repository; upgrading of nuclear reactors for producing defense nuclear materials; and a mammoth chemical weapons stockpile reduction program will typically take from seven to twelve years from start of engineering to operational readiness. Some elements of the Strategic Defense Initiative, or "Star Wars," program are expected to take up to twenty years from concept formulation to routine operation in space.

A dynamic, integrated, national planning process has been needed for many years. Until a few years ago, the thought of putting all of the government's major systems into an overall twenty-year plan would have been mind boggling! Now, however, with the advent of high-powered computers, excellent scheduling software, and highly capable computerized data base systems, an overall dynamic plan, with more detailed agency and service plans, is not only conceivable but is just a few years away from conclusive practicality. One of the first major initiatives to be entered into the twenty-year plan will be the development of the planning process itself. Sufficient cost/benefit and feasibility studies of such a process could be completed by 1989 to permit a national planning framework to be in place by 1990. Within five years, a computer-assisted dynamic planning network could be operating at a cost far less than the potential cost savings that could be realized through future use of such a plan. Major systems studies[8] have shown that many of the problems of sporadic support, inadequate funding, and overlapping projects that have occurred in the past could have been avoided if such a national plan to meet national goals had been formulated and approved in advance. But now, it is a concept whose time has come!

A viable, national long-range plan for major systems—developed as a basis for firmly approved, multiyear, full-phase funding—can help eliminate one of the major contributors to cost escalation: funding instability. A broad plan will also assist in the formation of political policy, and its dynamic nature will permit adaptation to technological evolution. These principal benefits of a dynamic national plan are to be gained through the visibility of the plan to the President and the Congress, and through the political interactions that occur in the planning process itself. A cost avoidance in major projects of up to 50 percent in selected programs is inherently feasible by eliminating sporadic start/stop or speed-up/slow-down costs. As reported in many of the studies we reviewed, multiyear, full-phase funding will permit (1) more economical production buys, (2) greater efficiency in the use of facilities and personnel, and (3) more consistent support and efficient deployment of the U.S. technological, academic, construction, and industrial bases.

The Phased Acquisition Process

Experience has shown that a time-phased acquisition process will maximize the three *e*'s (efficiency, economy, and effectiveness) in procuring major systems. The five or six major phases of acquisition have been given many names by many people, but for the purpose of this discussion we will call them Phase I: Program Initiation Phase; Phase II: Demonstration and Validation Phase; Phase III: Full Scale Development Phase; Phase IV: Production (or investment) Phase; Phase V: Deployment Phase; and Phase VI: Operational Phase. Two or more of these phases are sometimes combined to result in fewer phases, but their sequential nature always shows up in successful projects. In our interviews and in many of the reports reviewed on major systems, it was evident that concurrent rather than serial phases resulted in cost escalation, performance shortfalls, and eventual schedule slippages.

Despite the avowed dedication of government planners to life-cycle costing, programs are usually short-changed at the beginning because of the pressure to get the money flowing. For example, in the initial design of NASA's space station, automation and robotics were neglected to save early funding while their use would have decreased overall costs.[9] A NASA official said that "the constraint to design the space station to an initial cost target without giving due weight to operational or life-cycle costs is the primary inhibitor of automation and robotics for the station." He said that project managers are driven only by cutting initial design, development, test, and evaluation costs because there are no longer-term cost targets. If life-cycle macro-planning and macro-estimating of systems like our national space station were done within the context of a twenty-year dynamic national plan, decisions could be made based on life-cycle costs rather than just the cost for the first year. A change from the short-term view to the longer one will take a strong desire and dogged patience in the present world of quarterly profit statements, programmatic near-sightedness, and a "let's do it all this year" philosophy. But there is increasing evidence that a dynamic national plan, coupled with phased acquisition of major systems, will provide a focal point for rational and effective acquisition. Use of the phased acquisition cycle, if pursued with sufficient discipline and self-control, will let us commit to selected life-cycle cost-effective projects on a phase-by-phase basis, with each phase approval contingent upon the successful completion of the preceding phase. Phased acquisition with full-phase funding will provide the stability needed to eliminate diseconomies caused by start/stop funding.

The Dangers of Concurrency

A policy that has had, perhaps, a more detrimental effect on large programs than any other in terms of achieving desired performance or specifications within reasonable resource constraints, is the policy of concurrency, which for purposes of this discussion we define as parallel development and production. Total concurrency, or starting development and production at the same time, is the exact opposite of phased acquisition. Programs where development and production overlap involve varying degrees of concurrency, depending on the amount of overlap. We should avoid concurrency altogether if it is defined as parallel development and production of the same item. It is for this reason that we will devote considerable space to the discussion of concurrency and its variations. The concurrency approach to acquisition and its potentially disasterous effects, as foreseen many years ago by the Major Systems Acquisition Study Group of the Commission on Government Procurement[8] were described in the study group's final report as follows:

> *A family of problems arose from the importance attached to production as a justification for investment in system development. This, in part, was a conditioned response to persistent criticism of curtailment or cancellation of projects as wasteful and an admission of failure. To avoid waste and failure meant to be more selective about beginning projects and then* not *to cancel a development once begun, but rather to go on through production.*
>
> *Building on this premise, it became clear that the high road to cost reduction was through shortening the development time cycle. There seemed no better way to this end than the concurrency approach to acquisition. In this approach, development and production schedules overlapped and production was begun on the assumption that the development would be largely successful. Necessary and inevitable changes could be accomplished by introduction at periodic assembly line changeovers or by retrofit in the finished items.*
>
> *The evidence to support this cost/time trade-off was formidable, especially if one looked at the records of programs of highest national priority where there was, literally, no alternative to success.*
>
> *Problems came with the institutionalizing of this theory into procurement doctrine that decreed concurrency to be the "best of all." Total package procurement was then but a short step away.*

It is revealing to find that the concurrency policy was nonetheless an uneasy one, and that debates regarding its superiority over more cautious approaches were continuous and, on occasion, bitter. The misgivings, however, were not reflected in directives or in practice, and concurrency, with its spawn of "ilities," drove the more moderate forms of system acquisition out of circulation. It has been suggested that a form of Gresham's law—that bad money drives good money out of circulation—works effectively in the system acquisition process where tough procurement practices seem to drive more reasonable practice out of circulation.

It can now be said, with the advantages of hindsight, that the values placed on concurrency were exaggerated.

The purpose of a development program is to get ready for production. Concurrency is a paradox because it presumes startup of production before development is complete. This means that production will start before it is *ready* to start. The obvious result of concurrency is a constant change in the item being produced, which defies the definition of production (the replication of numerous *like* items). Cost benefits in production result from the economies of repetition. What is really being done in concurrent development and production is the construction of a series of prototypes, each different from its predecessor. Since prototypes typically cost up to twice as much as the first production units, the "production" ends up costing more than it would if a number of like units were manufactured in sequence. Not only does cost go up dramatically, but the dissimilar products that result have different operational characteristics, making it difficult to evaluate test results, establish field operations, or develop a statistical base to determine reliability or operational suitability.

The disastrous effects and painful results of concurrency were devastating to Adolf Hitler during World War II. He ordered mass production of the V-2 missile less than three months after the first successful prototype flight test on October 3, 1942. Ordway and Sharpe describe the effects of concurrency on the V-2 missile program in their 1982 book, *The Rocket Team*[10]:

Although Degenkolb [manager of component production operations], Stahlknecht [special commissioner in the munitions ministry], and others were thinking of the A4 [V-2 prototype] in terms of mass production, the missile still remained very much in the research and development stage and did not enjoy the top priority rating that Speer, Dornberger, and von Braun knew was indispensable . . . .

Although Degenkolb had clearly announced his ambitious production goal for the A4, engineers up at Peenemunde were not yet able to deliver a production-ready design to him . . . .

At one time or another, almost everyone involved with the A4 complained of troubles they were having in securing reliable components and in minimizing design changes. Stahlknecht lamented that preproduction design drawings "come in so late and are no sooner delivered than they are fundamentally changed again." Von Braun talked often of frustrations arising from attempting to accommodate the A4 to the continually changing raw materials situation. Not being able to rely on deliveries of a given set of materials "keeps on forcing us to make designs with other materials said to be in better supply." (At one time, virtually the only aluminum available came from Allied bombers that had been shot down.)

Eberhard Rees was dissatisfied with the quality of components received from the contractor and subcontractor network; rejects were common. Propulsion chief Thiel reported that "the components [that industry] does turn out are not in any way up to standard. The motor is too complicated and very far from suitable yet for mass production." In a mid-August meeting with General Dornberger, Thiel despaired: "For months now we have had one breakdown after another. We expected too much of our A4. In present conditions the job just can't be done. Our machine is a flying, fully automatic laboratory. To put it into mass production is sheer madness. We aren't through with the development by a long way . . . . If you persist in your point of view [to ready the missile for mass production] I must decline all further work. I see no possibility that we shall achieve our aim before the war is over. The project must be abandoned. I have given the whole matter thorough consideration and ask to be allowed to resign. I intend to join a technical college as lecturer in thermodynamics.[a]

The same German team that was forced into concurrency in the V-2 program passionately avoided it when they came to the United States to work on the Army's missile program and later the Apollo program. Thorough, meticulous, painstaking, systematic development backed up by an extensive laboratory and prototype testing program was required before operational flights could commence. The meaning and importance of the design freeze were well established, and full testing before making "block" changes in production became known as a way of life for a maximum chance for success.

Developers and manufacturers of commercial products as well as military systems have found that concurrency is to be avoided. Lee Iacocca described his unfortunate encounter with the results of concurrency just after he took command of the Chrysler Corporation in 1978.[11] Owners were complaining of engines that stalled, brakes that failed, fenders that rusted, and hoods that would fly open in the Dodge Aspen and Plymouth Volare automobiles. Iacocca said that these cars were introduced six months too early because the company was hungry for cash and that the company therefore did not follow the industry-accepted standard cycle for designing, testing, and building an automobile. Iacocca said, "The customers who bought Aspens and Volares in 1975 were actually acting as Chrysler's development engineers."

In addition to hardware products such as missiles and automobiles, automated, computer-driven systems must also avoid concurrency. For

[a] The quotation is provided from *The Rocket Team,* copyrighted in 1979 by Frederick I. Ordway and Mitchell R. Sharpe, through the courtesy of Harper & Row, Publishers, New York.

example, the General Accounting Office, in a report on the Federal Aviation Agency's "Advanced Automation System" acquisition strategy pointed out the following:

> *FFA's acquisition strategy is accelerated—it combines development, test, and production into one phase. Full production of the costliest element—the controller workstation—occurs concurrently with development of the most complex software and hardware elements. Typically, major acquisitions have a separate development and testing phase that precedes the commitment to full production. Production decisions are usually based on operational tests of actual performance. In contrast, FAA intends to make the production decision based on contractors' paper designs, computer-model simulations of system performance, and design trade-off analyses. No operational testing will be conducted prior to the production decision. We believe FAA's strategy has unacceptably high risks and may result in significant cost increases, schedule delays, and performance deficiencies.*

The GAO's recommendation for further testing prior to production was accepted by the FAA, but the FAA has decided to choose a system based on a preliminary design only rather than on a final design. This tells us that the system decisions will be based on testing of other than the final design configuration, which also could result in detrimental delays and cost increases in the program.

The results of accepting the high-risk concurrency approach for the B-1B bomber were described in a February 4, 1987, address to the National Press Club in Washington, D.C., by General Larry Skantze, Commander of the Air Force Systems Command. General Skantze said, "My point is that the problems we are seeing in the B-1B today are products of the risks we accepted when we decided to develop *and* [emphasis stated in context] produce the aircraft concurrently."

Frank C. Conahan, assistant Comptroller General, testified to Congress at about the same time[12] that meeting an October 1986 operational capability date for the B-1B bomber would "require a high degree of concurrency between development and production and, in fact *some development and production contracts were signed on the same day*" (emphasis added). As mentioned earlier, he concluded that a high degree of concurrency between development and production was a major contributor to the program's problems despite the fact that Air Force procurement regulations dictate generally that development, production, testing, and deployment of a major system be conducted sequentially.

Mr. David Packard[5] said, in his foreword to the Packard Commission Report, "If DOD truly is to fly and know the cost before it buys, the early phase of research and development must be one of surpassing quality, following procedures and meeting timetables *distinct from* those of approved production programs" (emphasis added).

Hence, concurrency is just an inappropriate concept for major systems acquisition which never should have been continued, particularly after the warnings of the Procurement Commission's study group and many other government, industry, and academic experts. The evidence is overwhelming that programs advocating true concurrency of development and production are headed for large cost increases and potential performance shortfalls. Even the *overlapping* of development and production can have harmful effects because often the last ounce of performance that spells success is not squeezed out of the development program until its very final phases. When this is the case, we end up with a large quantity of production items sitting in storage that just do not do the job.

There will be those who argue that concurrency is not bad, in and of itself, for every conceivable set of circumstances. They may argue that it is quite appropriate, in fact, for programs that involve no major technological uncertainties. The point they overlook is that even minor technological uncertainties require some research, development, or testing, and that these critical tasks need to start before production begins in order to assure maximum cost effectiveness in production through the replication of like systems or products. If there are no technological uncertainties at all, then, of course, production can begin. Product improvement can parallel production if the improvements are thoroughly tested in both developmental and operational environments prior to their introduction into the assembly line. Again, the basic design of the parent system must be pre-engineered to be receptive to modular add-on changes or improvements. Thorough testing and perfection of both the baseline system and any modifications must be completed before either is produced in quantity. Elimination of concurrency of development and production in both the baseline system and its upgrades will prevent expensive retrofitting of either to make them work.

Several programs, including the MLRS project described in Chapter 2, have been successful using a concurrent *purchasing policy* that they have called "concurrency." In this more restrictive definition, standard materials or parts for production units were bought at the same time as those for development or prototype units. This technique merely represents good planning, management, and stewardship of the taxpayers' dollars. It does not force the horse to run parallel to the cart as does concurrent development and production.

Concurrency, as defined here and as advocated by procurement innovators who are trying to speed up the whole acquisition process, should not be confused with the concept of *all-up-testing* used by NASA in the early Apollo days. In all-up-testing, the unmanned prototype missile or space vehicle is flown without complete subsystem testing in both simulated and unmanned flight environments. It saves time and money and is based on the premise that the flight vehicle itself is usually the best and cheapest test bed for its subsystems.

Concurrency should also not be confused with "evolutionary development," an approach used successfully by several agencies in which an initial conservative design is thoroughly tested, put into production, and then modified or upgraded through thoroughly tested block changes. The fact that development is proceeding on potential block changes while the baseline system is being produced has no detrimental effect on the production program so long as the block changes are as meticulously tested and proven prior to their introduction into production, *and* provided that the initial design was conservative enough to permit easy retrofit on already produced items. Examples of the use of this approach include providing a beefed-up structure to permit later performance increases and providing sufficient on-board computer capability, maneuvering capability, or payload to accommodate future software, control, or requirements for passengers, cargo, or weapons.

At about the same time the Procurement Commission was formulating its final recommendations to Congress, NASA was conducting a management study of its own acquisition process. In its 1971 final report,[13] NASA restated its position as follows:

> Recommendation: *That the basic objectives of the current project planning policy be reaffirmed, i.e., that projects should evolve through a series of logical steps from concept through development.*

In over thirty years of research in improving the military acquisition process, the Rand Corporation still concludes that sequential, not concurrent, acquisition is the way to procure major systems.[14] All of this tells us, then, that sequential or phased acquisition is the prudent basis for building an integrated, dynamic national plan for major systems.

What Does Industry Say?

Industry has a lot to say about the government's acquisition planning, and in 1971 we found that they were no less vocal on both an individual and collective basis than they are now. In the comprehensive major systems study for the Procurement Commission, we held private one-on-one or small group interviews with chief executive officers, vice presidents of engineering and finance, project managers, and contracts personnel from the defense and aerospace industries. We asked each of them, "What do you think are the major causes of the present problems in major systems procurement by the government?" The typical industry executive, manager, or contracts person responded like this:

> *First, industry has the perception that the government does not know how to estimate costs. The government needs (1) good people who really know what it takes in industry to develop and produce a product and (2) good cost-estimating methods that will produce accurate results.*

> *Second, there needs to be consistency, continuity, and direction in long-term plans and goals. Industry has the perception that the government doesn't know what it really wants and when it wants it. Ground rules change, programs change, and schedules change. Programs are frequently started, stopped, or delayed at inopportune times, creating huge cost increases. Once a program is started, we are infrequently provided with adequate funding on a regular basis.*

Government contractors were deeply frustrated. If they bid realistically, they did not get the job. If they bid low they "won" the job but, thereafter, had to battle for sufficient funds through the changes or "claims" route. There was at least one instance where we interviewed the participants on both sides of the negotiation table (the industry people and the government people): the negotiation between the government and Lockheed on the C-5 cargo aircraft total package procurement. The total package procurement concept combined development and production into one contract and one phase; thus, it represented the ultimate in concurrency. This program was one of the biggest procurement disappointments in the history of major systems acquisition; in fact, its failure to meet original commitments was one of the forces that caused the formation of the Procurement Commission. Negotiators on *both* sides of the table told us privately that they *did not realistically expect that the project could be done at the price being negotiated.* There were tremendous budget pressures to keep the cost down—the "total package procurement" method had been mandated by then-Secretary of Defense Robert McNamara. The negotiators told us that they were thinking, "Let's get on with the project anyhow—perhaps a miracle will happen and we will be able to produce an acceptable aircraft at the quoted price." But the extent of competitive optimism, and the desire by both parties to get the program going, overrode the better judgment of the Air Force procurement officials as well as the Lockheed negotiators. (It is a matter of record that this total package procurement did not work.[8])

The C-5 example was only one manifestation of the more basic and fundamental problems that existed in large systems procurement at that time—and which still exist, even after much revision, additions, reorganization, and "reform" in the acquisition process. It was also a graphic example of the underlying drives and frustrations that were revealed to us again and again in our person-to-person interviews with industry.

Much has already been done to correct the first concern of industry: the need for improved cost estimating in government. The emergence of cost estimating and cost analysis as recognized professions has helped to coalesce the skills, information, and methods needed to pinpoint more accurately the expected costs of a major project. A dynamic national program plan will provide a vehicle for the effective *use* of these cost estimates.

The second concern of industry, the inadequacy of planning and the absence of a road map for industry, still exists.

The Industrial Base

Any dynamic national program plan must take into consideration the overall policies of the U.S. government to employ industry in obtaining its major systems and projects. Studies on major systems acquisition have recognized the desire to establish, maintain, and support an "industrial base."[5,15] The military requires a "mobilization base." The scientific and academic well-being of our country requires a "technology base." Specific industries lobby for funding to support their continued participation in large federal projects. The construction, manufacturing, aerospace, electronics, and computer industries are also part of an overall U.S. "industrial base."

The industrial base significantly influences the policies used in acquiring major systems and should benefit from and have an input into the dynamic national planning process. Economic and political pressures put on the Congress by special interest groups such as small businesses, professional societies, and trade associations influence not only how systems are procured but what systems are procured and when. Use of existing workforces, economic development, and environmental issues raised by special interest groups have a bearing on the geographical location, magnitude, and occasionally even the *type* of systems and projects procured.

If one of the objectives of government procurement is to fashion and sustain an industrial and technological base, then industry must be a contributor to and a beneficiary of national major project planning. Participation in this planning will provide a basis for marketing, capital investments, and corporate strategic decisionmaking.

Forming and sustaining an industrial base capable of producing our *future* major system needs can be compatible with cost-effectiveness in the acquisition of *present* major systems, just as providing an adequate defense capability can be compatible with efficient acquisition of weapon systems. To do both, however, the major participants must have agreed-upon short-term and long-term goals. Can this be done and made to stick even beyond the term of a current administration? Yes! President Kennedy did it when he said, "Let us put a man on the moon in this decade!" He placed a target milestone on the national schedule that could serve as a rallying point for industry, government, and the people.

There is increasing pressure from industry, as well as from others in both the private and public sector, to develop a rational process whereby the Congress and the Administration can reach accord on a budget, a budget that fits within our financial resources and supports agreed-upon long-term goals.

Of necessity, a long-range dynamic national plan must have sufficient visibility to permit planners to fully and efficiently use the industrial base in acquiring its major systems. This visibility must be provided without compromising national security—a difficult but not impossible task. Presently the federal budget is an unclassified document available to any citizen on request or for purchase at a nominal fee. A dynamic national plan, formulated, modified, and updated in much the same manner as is the federal budget, could provide the same general resource category and geographical breakouts as provided by the federal budget and its supporting documents, but over a long enough period of time to cover major system and project developments. Since the plan would not contain technical specifications, designs, or development plans for specific systems—only overall schedules and budgets—there would be little more hazard of exposing our detailed plans to potential adversaries or competitors than there is now with the federal budget review, debate, and approval cycle.

The availability of longer range projections and the ability of industry to participate in the planning cycle as it does now in the yearly budget cycle by exerting pressures for specific projects, systems, and supporting services will enhance corporate long-range planning, provide assurance of continued economic development and growth, and give forward visibility to major shifts in needs and resulting projects to fill these needs. Let the debates start far ahead of the potential needs so that longer term stability and more consistent projections will be possible in both industry and government. The dynamic national plan, of which the current fiscal year's budget would be only a part, will be a rallying point, a focal point, and a guiding path for long-term financial, economic, and technological planning and commitment for a stronger and healthier industrial base that will engage in the development, production, and construction of our major systems and projects.

Why a "Dynamic" Plan?

The nature of U.S. industry is not static; it is dynamic. Effective planning is dynamic in nature. During the formative stages of a plan, a high degree of flexibility is needed to ensure that all alternatives are considered and that the best alternatives are chosen. As plans for specific systems become more firm, these systems must be *baselined,* and the baselines updated only when improvements can be accommodated within available budgets. Thus, a dynamic plan consists of plans for many systems, some of which are firmly established, some of which are in the early stages of planning, and some of which are in between. The use of a dynamic rather than a static national plan will permit the flexibility inherent in a democratic society without perturbing programs that have already been approved and are already underway. A dynamic plan is also needed to permit the matching of a given mix of major systems and projects to projected long-term budgets. We see a situation in which certain high priority systems and projects, pre-selected through the Presidential/Congressional planning cycle, are chosen to fit within conservatively achievable baseline budgets. Other projects, standing in the wings after sufficient pre-planning of their own, are injected into the national plan when income from taxes or economic growth is sufficient to accommodate them. Thus, the mix of systems or projects in the plan at any time can depend on available funds without cancelling out urgent programs when fund availability does not meet expectations. If fund income to the federal government does not meet expectations or projections, the pre-selected lower priority projects would be put on hold until the funds are available.

Admittedly, this rigorous planning process places a great burden on those responsible for it—but so does the formulation of a multibillion-dollar major systems budget every year by the executive branches and agencies of the federal government. Lengthening the period of time, or the "window," for which planning and approval of projects takes place from one, two, or five years to fifteen or twenty years will serve to increase stability of programs as well as facilitate industry in planning its major capital investments, business decisions and directions, employment policies, and growth plans.

The National Attention Span

The need for a dynamic long-term plan for major systems is accentuated by what has been called a too-short national attention span. Approved long-term goals and plans are needed to foster the patience and persistence required to provide full programs or full-phase support for ten- to fifteen-year programs. National attention tends to focus on the initiatives or problems of the moment. Since it is the nature of the major system or project to require thoughtful conception and sustained nurturing for many years, a philosophy and a supporting mechanism must be available to bridge the gaps in the national attention span. Major changes in direction should be made at logical program phasing points rather than at the beginning or end of every government fiscal year. Full-phase funding will bridge the fiscal year successions, and full-term program planning must be credible enough to bridge transitions in Congressional and Presidential leadership. A *dynamic* plan to permit new Congressional or adminstration guidance must be assimilated at predefined program phase intervals rather than every fiscal year. The pre-planning, debate, analysis, and pre-commitment required to approve each program phase will heighten the awareness of the Congress of the long-term impacts of its decisions. Policy makers must be provided with credible answers to the questions: "What does this expenditure mean downstream?" "What are the long-range impacts?" and "What are we buying into?"

Virtually every report we have seen on major systems acquisition has emphasized the need for in-depth research, analysis, simulation, planning, and estimating *before* committing to proceed with a program or program phase. Since a dynamic national plan will be the result of greater in-depth planning, analysis, and debate *before the programs or program phases start*, the nation's attention will be placed at the front end of the process where it belongs.

In a 1971 presentation to the members of the Procurement Commission by members of the Major Systems Acquisition Study Group, the group made specific recommendations on "Major Systems Planning and Budgeting in Support of National Goals."[16] One area addressed was "the need for a broad, visible, long-range plan." The group found that the "microscopic" detailed planning of major systems was not matched by "macroscopic," or large-scale, planning. A twenty-year plan was proposed, which would encompass the major systems needed to meet long-term national objectives. The group observed that "an element of assurance can be interjected into an inherently uncertain process through thoughtful, judicious, and forward-looking planning" and that "the traditional elements of cost, schedule, and performance must be accompanied by a fourth factor: How (or *if*) a system fits into the matrix of evolving and changing national goals." The potential benefits of a broad, visible plan were listed as (1) provides a clear projection of the most likely path, (2) provides a baseline which can be updated as necessary, (3) provides a guideline for budget formulation, (4) provides a base for developing priorities, (5) serves as a guideline for industrial marketing, and (6) serves as a focal point for resource allocation.

In addition to these critical points, the dynamic nature of the plan, and of the planning process itself, will accommodate accelerating technological change and will provide a visible forum for debate, discussion, and input by inherently contending political interests. The need for a dynamic long-range plan is no less now than it was in 1971. And there is increasing evidence that the adoption of a long-range plan for major systems will help avoid one of the most significant controllable causes for program cost growth—funding instability!

Needed: Long-Term Goals

Little has ever been done without first setting goals. A dynamic national program plan will require agreed-upon national goals or objectives and targets or aims that represent desirable common needs, such as a balanced armed force or a revitalized interstate highway network. Many individual agencies, departments, and services are indeed striving toward imaginative individual long-term goals for major systems, but these goals have not been given top-level exposure and consideration by the Congress and the President as national priorities. Too much time has been spent on arguing about the next fiscal year, and too little attention has been given to the necessary synthesis of far-reaching objectives for both civil and defense systems.

Those of the 45 million people still around who attended the 1939/1940 New York World's Fair held in Flushing Meadow Park, Queens vividly recall "The World of Tomorrow" exhibit with its glistening cities; orderly, intertwining highway networks; moving sidewalks; self-guiding autos; rapid communications; and automated factories. The Fair captured the imagination of millions of people. Even those who did not personally visit the Fair heard of it in glowing news reports on the radio and in the newspapers. Scientific marvels of the future such as television, nylon, and air conditioning were promoted and displayed to the public enmasse for the first time. The Fair created excitement and anticipation among those who saw it and heard about it because it gave a glimpse of future megasystems and megaprojects that would improve their quality of life beyond anything they had previously imagined. There have been many world's fairs and expositions since that time, but none have quite matched the one in Flushing Meadow Park in its ability to rally a nation around a vision of the future. The fair was a goal setter, an imagination stimulator, a prototype of what was to come. We see a set of national goals as doing the same thing. As we approach the year 2000, where is our world of tomorrow? What will this nation be like twenty years (or fifty years) from now? Are we taking steps toward a new and more perfect society? If so, what do science and technology, industry, and government have in store for us? And do the major systems we are now developing fit into the future plans?

Goal setting is needed not only in civil systems—as manifested in the first New York World's Fair—but in defense systems. Where are we going with our Army, Navy, Air Force, and Marines? What new tools will they have available in the next fifteen to twenty years? Do the President and the Congress have a long-term plan for force modernization on land and at sea as well as in space? Are goals being set to provide our defense personnel with the tools needed to ensure our continued liberty? What long-term objectives have been set for the military? Mobility? Entrenchment? Readiness? Retaliatory power? Automated battlefield management? Targets need to be set up on the firing range of the future, as well as on the battlefield, to permit methodical planning, scheduling, and estimating of the resources, personnel, and equipment needed on a long-term basis, not just for the next five years. Successful megaprojects have shown that commitment and unswerving dedication over ten to fifteen years are required for efficient acquisition. Lack of this endurance has caused projects to falter and finally run aground.

A 1986 report by the General Accounting Office on the *Status of Open Recommendations on Improving Operations of Federal Departments and Agencies*[4] gave some clues to the urgent need for improved goal setting, detailed planning, requirements analyses, and a government-wide perspective in planning. Findings of numerous studies listed in this report have a bearing on the acquisition of major systems:

Treasury Department Computer Systems: "The absence of top management and user involvement and participation in formulating a *long-range computer growth strategy* has resulted in Treasury bureaus having either too much or too little computer capacity, excessive costs of operations, and unmet user needs."

Federal Aviation Administration (FAA) Computer Systems: "FAA has not made a comprehensive analysis of its overall information requirements. . . FAA does not conduct overall planning to meet agency-wide needs. Therefore, overlapping or duplicative systems have not been identified, long term planning has been impeded, and evaluation of the overall effectiveness of existing information systems has been hampered. Furthermore, information requirements analyses are not adequately conducted to support computer acquisitions."

All Federal Agencies—Computer Software Systems: "*No overall process* exists to ensure that Federal agencies consider alternative methods of satisfying software needs." [In the data processing installations that GAO visited, over 98 percent of the software inventories had been custom developed, which is a long and costly process as opposed to using off-the-shelf software.]

Office of Management and Budget (OMB) Telecommunications Initiatives: "OMB is planning to permit agencies to acquire long distance telecommunications services independently instead of using the centralized Federal Telecommunications System . . . . It is not placing enough emphasis on the *government-wide issues* arising from the new telecommunications environment." [The report states that OMB had failed to examine the long-term cost benefits of a centralized system.]

Patent and Trademark Office (PTO) Automated Retrieval System: PTO established major goals in 1982 but did not "(1) *thoroughly analyze* or develop the functional requirements for the use of its three automated systems; (2) *adequately assess* the costs and benefits of the automated systems; (3) *properly manage* the three systems; and (4) *fully test* the search and retrieval system before accepting it from the contractor."

Social Security Administration (SSA) Computer Systems: Software standards were incomplete, emphasis was shifted from improving existing systems as originally planned to building new systems, the new systems involved a technology beyond the current state of the art, and "*there was a lack of effective planning, control, and monitoring* of system projects."

Decommissioning Nuclear Facilities: The Nuclear Regulatory Commission and the Department of Defense have not "initiated action to develop a comprehensive decommissioning policy." Absence of this policy and a detailed plan has resulted in a lack of "consistent flow of funding."

Nuclear Waste Disposal: The Department of Energy (DOE) "has delayed many actions required by the [Nuclear Waste Policy] act because of unrealistic scheduling and inadequate contingency planning."

Office of Science and Technology Policy (OSTP): "It seldom studies the relationships of issues in the whole context of science and technology in society; instead, it usually focuses on a particular mission issue in isolation from its interaction with other national concerns." The GAO further states its beliefs that this office should be involved in "*comprehensive strategic planning*, and "that, within existing constraints, OSTP can establish a systematic and formal mechanism for identifying long range emerging issues and for providing a detached perspective in screening outside proposals."

As can be seen by these GAO findings, time after time, failures in meeting needs have been traced to a failure to develop realistic long-range goals, plans, and cost estimates both at the agency level and at the overall federal level. The indispensability of early user involvement in the goal-setting and planning process is also clearly indicated in these and other studies of large government projects.

An excellent example of how to effectively establish long-term goals and plans was a study completed in 1986 by a Congressionally legislated and Presidentially appointed National Commission on Space. The President appointed fifteen outstanding citizens from government, industry, and the academic community to develop a recommended long-term space program. The commission studied and selected goals for the next fifty years. The study was headed by Dr. Thomas O. Payne, a former administrator of NASA. The commissioners included current and former astronauts (Kathy Sullivan and Neil Armstrong); scientists, including a Nobel laureate physicist; retired Air Force generals (Bernard Schriever and William Fitch); industry leaders; consultants; educators; a test pilot (General Chuck Yeager); and former Ambassador to the United Nations Jeanne Kirkpatrick.

In a one-year program of public forums, interviews, workshops, and hearings, the commission gathered information and synthesized a national major systems program plan that extends to the year 2030. The commission report, titled *Pioneering the Space Frontier*,[17] contains concepts of space systems, a milestone chart of significant goals for the next 50 years, a preliminary estimate of the cost of the program, and a description of the program's objectives and potential benefits. Through the mechanisms of forums and public hearings, the commission went directly to the public to get inputs and comments. It solicited letters and conducted surveys to gather public opinion and sought advice from experts in space science and space systems. A commission staff collected and digested the information.

The commissioners met once a month to discuss the emerging conclusions and recommendations in detail. A total of 1800 people participated in public forums held in Los Angeles, Seattle, Houston, Salt Lake City, Tallahassee, Ann Arbor, Iowa City, Albuquerque, Boulder, San Francisco, Cleveland, Huntsville, Boston, and the District of Columbia.

Witnesses at the forums consisted of invited individuals, those who contacted the commission prior to a specific forum requesting time to speak, and unscheduled participants who made use of an open microphone. A large percentage of the people who attended actually made presentations. Disciplines represented by those who attended the forums included theology, philosophy, and teaching (elementary, high school, and college levels). Former astronauts, folk singers, homemakers, lawyers, engineers, and Congressional leaders also attended. The bulk of those attending the forums had no direct link to the space program; in other words, they represented a cross section of the interested general public. The commission said, in its final report, that "We were overwhelmed by the high calibre of comments obtained and duly impressed by the commitment of the citizens in attendance to respond intellectually to the call for participation."

Since it was a fifty-year plan, there was great emphasis placed on the younger generation, the ones who would have to pay for—and be the principal future beneficiaries of—the long-range program. Emphasis was placed on finding ways of educating the future generation for the building and use of future systems and projects. (Any such long-term plan would necessarily include the younger generation and those who are charged with their education and training.) One teacher said, "Now we are seeing a new frontier, a frontier that belongs to our young generation. How are we going to educate those children if we, the older generation, don't understand what's going on out there?" Continuing, she remarked, "We need to have a bigger vision."

The commission members were repeatedly thanked by citizens for creating the public forum concept and coming to their city. "In essence, our activity was viewed as bringing the space program to the people."[17] The several themes that were brought forward repeatedly from the forums included the following:

- The public has a desire to learn more regarding the scope and direction of the program and to assist in shaping its fate. A perceived lack of information on current and projected goals was heard throughout the forums.
- A cautionary sentiment was expressed that the commission's final report not be too conservative, and that it not highlight "one-item gimmicks," but promote the entire integrated program.
- There was tremendous support for the role of international cooperation in space.

- Incentives to strengthen private sector involvement were expressed on a number of occasions.
- A shoring up of the U.S. educational system and our ability to conduct research and development is urgently required "if the country is to truly move outward into space and remain competitive on earth."

Think of it! These people were not politicians, pounding the pavement for votes. They were scientists, administrators, engineers, academicians, school children, and people from all walks of life expressing their desires to a Presidentially appointed commission on where our country should be going in a specific field—space—in the future. The commission's final report listed the names of the people who participated in the forums, responded to questionnaires, wrote letters to the commission, and served as witnesses and workshop participants.

The final report, an attractive, 212-page book produced for the commission by Bantam Books, is a colorfully illustrated, easily digested synopsis of the study results. Fifteen thousand copies were given away to interested parties, and another 35,000 were made available for purchase. The report contained a listing of 150 other books, periodicals, articles, and reports that are available for more in-depth study. A photograph of each commissioner and his or her signature was included in the report. After the report was given to the President, he requested that NASA produce a detailed implementation plan which will show the specific systems, projects, and budgets required to carry out the long-term program.

The Commission on Space report represented an excellent way to develop and articulate national goals. Similar studies and similar reports could be developed for such fields as air transportation, ground transportation, communications, inner city development, and, yes, even our defense program. With goals established in all major fields of government sponsorship, a detailed, integrated, dynamic national program plan could be developed that would compile and prioritize the schedules for developing and producing major systems to meet these goals.

Planning and Scheduling to Meet National Goals

In *The Art of Japanese Management*,[18] Richard Pascale and Anthony Athos say that Japanese management is different from American management in its emphasis on several of the seven *S*'s (strategy, structure, systems, staff, skills, style, and superordinate goals). When an American manager wants to make changes, emphasis is usually placed on modifying one or more of the first three: strategy, structure, or systems. The tremendous success of many Japanese companies, the book says, comes through meticulous attention to the "soft" *S*'s: staff, skills, style, and superordinate goals. In any organization in government or industry, the most important and first step, as we have pointed out, is to establish superordinate, long-term goals. The next step is planning. Time-based planning involves a discipline called *scheduling* (another *S*). The success of the Japanese in carrying out large projects has been due, in a great part, to their expertise in scheduling, which contains three more *S*'s: synthesis, synergy, and sequencing. Synthesis is the ability to assemble and activate the people, organizations, equipment, facilities, and resources to accomplish a large task. Synergy is the ability to take maximum advantage of the interactions between the elements of a project. It is the attribute that produces an output or result that is often greater, better, or more useful than the sum of the inputs. When synthesis and synergy are accomplished on a time-based activity such as a major system or project, the third factor, sequencing, comes into play. The Japanese are masters at the art of sequencing activities to bring about the most desirable end project results. Through ingenious combination of synthesis, synergy, and sequencing, they take full advantage of the economies and efficiencies that result from expert scheduling. This is a key lesson that we can learn from the Japanese in the detailed planning of our major systems.

Tools and methods are now available and are emerging from ongoing research that will make it relatively easy to develop a detailed, dynamic national program plan for major systems acquisition. Whole families of project management computer software that will help plan, schedule, estimate, integrate, and track large numbers of large complex projects are already available. The Project Management Institute published a list of over 100 software packages that do planning, scheduling, and estimating of large numbers of small and/or large projects[19] with various types of computers. These software packages can schedule up to 500 projects at a time, each containing up to 2,000 jobs and over 4,000 resource allocations. Projects can be scheduled by year, month, week, or even daily if desired; resource estimates can be input for each job, and total resource requirements can be automatically compared with resource availability. Quick changes can be made in the sequencing, funding level, and milestones within projects to develop a schedule that results in the optimum use of available resources. Many of the available scheduling programs can display graphs of resource needs, and some can compare estimates with actual progress. Several software packages will permit the consolidation of projects from a lower level into a master- or top-level calendar-based bar chart. Critical path methods and resource-shaping methods are used to permit schedules to be adjusted for the most efficient sequencing.

The increasing use of telecommunications to transmit data, coupled with the emergence of excellent computer scheduling software, will soon make it feasible for agencies and their subdivisions to produce integrated top-level and detailed schedules for internal planning and to transmit their updated top-level schedules to central focal points like the Office of Management and Budget (OMB) to assist in the preparation of multiyear budgets to meet updated goals of the dynamic national program plan. Rapidly advancing technology that is expanding the speed, memory, and storage capabilities of computers, and continuing developments in computer software, including artificial intelligence, will provide the high-tech tools needed to effectively, efficiently, and economically plan and carry out our major systems of the future. Any study and synthesis of national planning requirements must include the definition of these government-wide automated systems and tools that will be needed to help existing organizations carry out the dynamic, long-term, in-depth planning function.

Comprehensive strategic planning must be used to design and to perfect the planning systems and methods themselves. One future national goal will undoubtedly be to improve our long-term planning methods and tools, and to put them to use in acquiring major systems. Planning and scheduling of the tools themselves will require the application of synthesis, synergy, and sequencing.

National Systems Planning

To develop a set of national goals, and detailed plans of major systems to meet these goals, it will be essential to have and to keep everyone involved. *Straw-man* goals and schedules will have to be developed by the various government agencies as starting points for the review, analysis, and public debate that will ensue. Obviously, all plans will not be affordable at once. But if goals and plans are based on current budget limits, we may be surprised how well they fit together—even initially. This strawman national goal structure and systems plan would be a focal point for discussion and a basis for debate and inputs to the planning process, a way of getting people to think about the long-range implications of projects, their interactions, and their potential benefits. We see the media playing a large part in publicizing the long-term plan and the debates that will inevitably result from an exposition of national priorities implied in the plan. Political parties may choose to formulate their own versions of the national goals and plans for exposure and discussion in the press and throughout the political process. Lobbyists and special interest groups should be encouraged to provide their inputs. Each should develop long-term goals and plans for its industry, profession, skill, or trade group. Each, of course, will be looking for synergy with other groups to broaden the political base of support for its special interests. The planning process will thus find ways of turning the self-serving interests of special groups into national interests. This concept will put these groups to work in developing positive, futuristic, forward-looking proposals, and they can become helpers instead of adversaries in the planning process. The objective of the national goals and planning process will be to emphasize the long-term benefits of integrated and far-reaching major systems and programs rather than merely their short-term gains.

How could such a process be carried out? What ground rules must be established first? And what procedures would be used?

Important sources of planning input and focus are the almost 3 million U.S. Government employees and the agencies themselves. Using the "centers of excellence" approach advocated by Mr. Packard, the organization can solicit ideas from all levels. Innovation and imagination must be encouraged. So far, there has not been an adequate incentive for this to work consistently and on a widespread basis in government; leadership is required that will rally national support and the dedication of government employees to be as supportive of innovation as those in private industry. Ideas should be solicited on how the federal government can be reimbursed for funds spent on major systems as well as how the government should *spend* money on major systems. For example, the government now recovers administrative costs from the sale of major systems to foreign governments through arms sales and charges user fees to foreign governments or commercial industry for transportation systems such as ships, aircraft, and the space shuttle. Above all, the goals and planning process *must not* become just another bureaucratic requirement that becomes cumbersome and meaningless and is done at the last minute just to get out a report. Time and attention must be devoted to the process at all levels.

We held a brief "think-tank" session to develop an outline of some thoughts and inputs as to how national goals for major systems could be established and how a dynamic national plan could be developed that would meet these goals. We assumed, for reasons stated in Chapter 1, that *existing organizations* would be used to carry out the national goals and planning process. The following outline was derived from our think-tank session:

I. How to Establish National Goals:
  a. Require each agency, department, or service to provide specific long-term goals (fifteen to twenty years).
  b. Update goals as necessary once per year.
  c. Methods of obtaining inputs for goal selection:
    1. Obtain inputs from industry, government, and the academic community.
    2. Hold public forums with interested citizens in diverse locations in the United States.
    3. Conduct computer-based polls among the experts in the field.
    4. Establish a hotline for innovative inputs and ideas.

5. Encourage letters from the public and interested parties.
6. Statistically analyze and synthesize the results of the above to develop trends that reflect the consensus.

d. Publish a goals report and make it available to the public periodically (this report should show previous goals, new goals, and reasons for the differences).

e. Through debate and the political process, the President should propose and the Congress adopt a national goals structure.

II. How to Produce a Dynamic National Program Plan

a. Each agency, department, or service will develop a detailed implementation plan to accommodate the approved goal structure. This plan will show each major system, its phases, and proposed new major systems to fulfill the long-term plan. Each plan will show the funding for the past five years and the next fifteen years. Program funding is broken out by program phase.

b. Proposed long-term agency plans are submitted to the President. The President, with the help of OMB and with the oversight of Congress, establishes priorities and publishes an integrated twenty-year plan.

III. How to Establish Fiscal Year Budgeting within the National Plan

a. The executive branch, using the approved national plan, develops the fiscal year budget for major systems. Approved ongoing phases must be funded. Phases that end during the fiscal year along with new program starts are weighed against the national goals and plans to provide continuation and/or new starts of the top priority programs that can fit within the budget.

b. The budget preparation cycle for each year continues in an iterative process, as it does now, until it is approved by Congress.

IV. Where the Resources Come from for National Goal Setting, Planning, and Budgeting

a. Use time saved by high-tech tools such as computers.

b. Use inputs from lobbying groups, trade societies, professional societies, special interest groups, and others to assist in the goal-setting and planning process. Use co-located personnel loaned from these organizations.

c. Use high-capability planning and scheduling programs to develop and update the dynamic plan.

d. Use standardized formats and electronic transfer of data to integrate plans as they pass from the lowest level to the highest level in the hierarchy.

e. Use the latest automation techniques to compare resource availability with collective needs, to pinpoint trouble spots in the budget, and to select programs that fit within the budget.

V. How Do We Keep This from Becoming Just Another "-ility" or Troublesome Report Required?
   a. Motivation from the top is essential. Make it the second highest national priority (the first is a standardized government-wide accounting system).
   b. Scenario: Our agency is required to have a long-range systems acquisition plan six months from now. We are to set up a series of meetings (think-tanks) to begin formulating our plan for our agency and are to provide the President with a feedback from each session so that, when the deadline comes, an iterated plan will be available to the President and to the Congress.
   c. This must be a creative process. It must be based on the "centers of excellence" approach. It must be the result of free, unrestrained thinking. It must be based on inputs from lots of sources—industry, suppliers, consultants, the academic community, the general public, users, and special interest groups.
   d. We must give each organization and each person an opportunity to be creative, to provide ideas and suggestions even though the ideas don't fall into his or her responsibility. The information society opens a way to creativity and participation that we did not have in the industrial society; this participation provides higher job satisfaction and permits real contributions by everyone. No suggestion or idea should be immediately rejected; all inputs should be considered as viable ideas and should be given thorough consideration. These are the same principles used in the quality circle approach and in participatory management.

VI. Remember that ideas and innovations are not just limited to those that *spend* money but can include those that *return* money.

James C. Miller, director of the Office of Management and Budget, said in his keynote address to the Fifteenth Annual Financial Management Conference of the Joint Financial Management Improvement Program,[20]

> *It's my belief that the creation of a budget is part of a management process. For too long, creation of the budget has been a narrowly defined end in itself. No more. The budget process should be an exercise that forces the right questions to be asked and answered. It should be* one of several important management plans *that are prepared each year [emphasis added].*

Mr. Miller went on to say, "The President must ask the same questions that every effective Chief Executive Officer (CEO) of a company has to ask: What is my company here for? What is our mission? What are our goals?" *What are our short-term/long-term business plans?* What are our implementing strategies? The budget process must be an integral part of the overall planning process. Every dollar in the budget must help implement one or more long-term goals. The present year funding and objectives must be compatible with the long-term plan's funding and objectives. One of the several important management plans mentioned by Mr. Miller should be a dynamic national program plan. The same iterative process used in establishing priorities in the budget for a given fiscal year can also be used to establish priorities for a longer term plan and its projected budgetary requirements. There cannot be adequate *management or budgeting* without long-range planning. The synergy and synthesis between budgeting and planning are axiomatic. A well-thought-out, iterated, long-term plan for major systems will allow the executive branch, with the oversight of Congress, to establish full-phase multiyear funding for approved major systems, thereby effectively treating one of the major causes of inefficiency in major projects, funding instability. Since the executive branch prepares the federal budget every year, it can provide a track record of budgeted versus actual expenditures, produce credible budgetary estimates for future years, and keep track of major reprogramming actions.

To do this mammoth job, top levels of the executive branch will need *much help* from the agencies and departments within the executive branch of government. Several years ago the current administration initiated a program called "Reform 88," in which agencies were directed to improve management, develop efficient and compatible administrative systems for the entire federal government, and improve resource management. Many of the initiatives of Reform 88[4,21–23] must be carried forward into the next administration and into the following administrations if the United States is to achieve a "corporate memory" that will produce lasting improvements based on proven success formulas. To do its job effectively, the government will need to have an effective government-wide management accounting system (discussed in Chapter 4). Detailed goals and plans will have to be prepared and periodically updated in compatible format for input to the long-term fiscal plan. Major automated financial systems based on new accounting formats that will permit electronic transfer and updating of schedule and cost information must be developed in the next decade to accommodate the complexities of a dynamic national implementation plan. In a country of huge corporations, the same techniques that run businesses will have to be applied to the federal government.

Several people who reviewed this book before its publication were curious as to how a dynamic long-range planning process could be put into effect. Some viewed it as something that would be done somewhere outside the existing government bureaucracy, perhaps with a new organization, department, or agency. This is a prevalent attitude that exists every time someone presents some "new" idea that appears to be an added responsibility in government: that it will require a whole reorganization, more people, more facilities, and more computers. Each organization that might be affected rushes in to say that such a requirement cannot possibly be done without hiring more people; getting improvements in salary structure for those who are already in the organization; adding new departments, equipment, or computers; or taking over some other organization.

For once, we would like to see the government take on an added responsibility without adding to the bureaucracy. This is particularly true because what we are recommending is (1) something that the government should be doing already without added push from the citizenry, (2) merely an extension in time from the *already* existing five-year planning to cover the time period now required for major programs, and (3) no more complex a job than is already done in the formulation and iteration of the yearly federal budget. The use of existing organizations, people, and computers could be accompanied by simplification and reduction of the microscopic detail now used in short-term planning in favor of a bigger picture for a longer time. We are firmly convinced that there is enough leeway in existing organizations to accommodate the formulation, maintenance, iterations, and exposure and discussion of such a plan with the Congress and, where appropriate, with the public. The plan would work best when formulated within existing organizations rather than imposed by another super-department or outside agency because planning personnel are already active in most organizations. Plans of existing agencies would be generated and iterated within the agencies just as they are now, and then fed into the cabinets, executive office of the President, and the Congress for synthesis.

Summary

A dynamic national program plan is needed to form a basis for multiyear budgeting, to ensure that major systems are being acquired in keeping with national goals, to optimize program interactions (such as in the space program, energy programs, research, and defense), and to provide a basis for selectively weeding out programs as time progresses. This planning process will help establish priorities and will avoid unrealistic expectations among the potential participants. The dynamic national plan will provide a basis for industrial planning and marketing and will provide a focal point for stability as well as adequacy of the industrial base. The current and continuing length of major projects—ten to fifteen years—makes it necessary to develop a plan that spans at least twenty years (fifteen years into the future and five years into the past). The past five years will always be important because they will highlight the investment that has already been made—an important input to any management decision. A dynamic national program plan will ensure the effective use of existing government organizations, ensure effective use of industrial resources, form a foundation for a continuing mobilization potential, and form a basis for continued industrial growth and development. Dynamic national program planning will be a systematic, methodical process designed to supplement and enhance—not replace—the present political interactive method of policy formulation and implementation. Decisions can then be viewed based on their overall impact on goals and long-term plans rather than the exigencies of the moment.

From Chapter 2 it is evident that efficient projects do, indeed, exist in the federal government. Hence, efficiency and democracy are not incompatible, as some would have us believe. Government procurement studies have shown that fundamental problems have arisen because of a lack of adequate planning and analysis prior to starting a program or a program phase. Foreign governments that have more limited resources than the United States have found that multiyear planning and budgeting are necessary in order to provide continuity of support and a smooth flow of funds to high priority civil and defense projects.

Development of a dynamic national long-range plan can complement the participatory form of government and can serve as a vehicle for the interaction of interested and chartered organizations in the process of providing our major systems. Planning focuses the participation and interaction at the *front end of the process,* where the most important and most potentially valuable trade-offs are made. Thorough, methodical, and systematic planning, analysis, and scheduling will give our projects the greatest opportunity for success.

To summarize this chapter, we must answer some of the questions most likely to be asked about a proposed dynamic national long-range planning process. First, how does the plan relate to the regular planning process in the executive agencies and the budget process? The answer to this is very simple: It must be an integral part of the current process. The principal change being recommended is not who will prepare the plan but what time frame is included in the plan. We have shown that megasystems take ten to twenty years to acquire and operate through their life cycle rather than two to five years. Clearly, a two- to five-year process is just not long enough to cover even the shortest phases of many major programs. No new organizations will be required to do dynamic national planning—only new directions to existing organizations. Second, how can major systems be considered separate from other planning and resources for the particular function involved? Each organization must take into account its long-term, short-term, and continuing institutional programs as well as its major systems, just as it does now in its shorter term budgeting and planning cycles. The long-term perspective for major programs must be integrated into ongoing activities. Again, it is the perspective that would be changed, rather than the organization or procedures. The longer term projects must be given high visibility; but we are not proposing separate planning, budgeting, and funding of major systems apart from the rest of the government's activities. Each agency must retain responsibility for its total mix of programs, systems, and projects. Third, who is going to be responsible within the executive branch for administering the national plan? This function must be managed at the highest level of government, which is the office of the President. We believe the public will come to expect each President to lay out the plan for all to see. Citizens will expect to be kept up to date on its progress, directions, changes, and status. This viewpoint will help the President keep focus on long-term directions for the nation in implementing short-term improvements and preventive measures. It will help the President see the impacts of current planning decisions on future administrations and will provide a focus for Congressional and public oversight, support, and approval of long-term major commitments for major systems and projects in the context of the total federal budgets for a score of years.

Endnotes

1. David I. Cleland and Harold Kerzner: *The Best Managed Projects,* Procurement Associates, Inc., Covina, CA, April 1986, GCS 8-86 D-1.
2. Lewis R. Ireland: *Project Management: Critical Success Factors and Keys to Effectiveness,* SWL, Inc., McLean, VA, August 1986.
3. General Accounting Office: *Status of the Defense Acquisition Improvement Program's 33 Initiatives,* General Accounting Office, Washington, DC, September 1986, NSIAD-86-178BR.
4. General Accounting Office: *Improving Operations of Federal Departments and Agencies,* General Accounting Office, Washington, DC, February 1986, OIRM-86-1.
5. Packard Commission: *A Quest for Excellence: Final Report of Packard Commission, 2 vol,* Packard Commission, Washington, DC, June 1986.
6. Large Scale Programs Institute (LSPI): *Colloquium on Research Priorities for Large Scale Programs 1985-1,* Large Scale Programs Institute, Austin, TX, March 1985.
7. Interview with Richard A. Reynolds, Director of Defense Sciences: Defense Advanced Research Project Agency (DARPA), Washington, DC, September 15, 1986.
8. Commission on Government Procurement: *Major Systems Acquisition—Final Report,* Commission on Government Procurement, Washington, DC, January 1972.
9. Martin Burkey: "NASA Faulted for Space Station Automation Plans," *Huntsville Times,* Huntsville, AL, November 16, 1986.
10. Frederick I. Ordway and Mitchell Sharpe: *The Rocket Team,* MIT Press, Cambridge, MA, 1982.
11. Lee Iacocca and William Novak: *Iacocca: An Autobiography,* Bantam Books, New York, 1984.
12. Frank C. Conahan: *The B-1B Aircraft Program* (testimony to Armed Services Committee), General Accounting Office, Washington, DC, February 1987, T-NSIAD-87-4A.
13. National Aeronautic and Space Administration (NASA): *Management Study of*

NASA Acquisition Process: Report of Steering Group, National Aeronautic and Space Administration, Washington, DC, June 1971.

14. Michael Rich and Edmund Dews with C. L. Batten, Jr.: *Improving the Military Acquisition Process,* Rand Corporation, Santa Monica, CA, February 1986, R-3373-AF/FC.
15. Commission on Government Procurement: *Listing of Study Topics from Approved Study Group Work Plans,* Commission on Government Procurement, Washington, DC, August 1971.
16. Rodney D. Stewart: *Major Systems Planning and Budgeting in Support of National Goals,* National Administration Association, Huntsville, AL, 1972.
17. National Commission on Space: *Pioneering the Space Frontier,* Bantam Books, New York, May 1986.
18. Richard T. Pascale and Anthony G. Athos: *The Art of Japanese Management,* Warner Books, New York, 1981.
19. Francis M. Webster: *Survey of Project Management Software Packages,* Project Management Institute, Drexel Hill, PA, October 1985.
20. Joint Financial Management Improvement Program (JFMIP): *Proceedings of the Fifteenth Annual Financial Management Conference 1986,* Joint Financial Management Improvement Program, Washington, DC, 1986.
21. David D. Acker: *"Reform 88": A Program to Improve Government Operations, Program Manager,* U.S. Government Printing Office, Washington, DC, March-April, 1983.
22. Office of Management and Budget: *Executive Order 12352 on Federal Procurement Reforms: A Reform '88 Program,* Office of Management and Budget, Washington, DC, February 1984, M-84-7.
23. Office of Management and Budget: *Budget of the United States Government FY87, 4 vols,* Office of Management and Budget, Washington, DC, 1986.

CHAPTER FOUR

Establishing Financial Stability

Undoubtedly the most important national initiative to improve the cost effectiveness of major systems acquisitions is the establishment of financial stability. By financial stability we mean: (1) the ability of a program to command continued as well as initial financial support; and (2) the ability of a program to stay within the bounds of original cost estimates. Because these two factors are interrelated, we group them under the single heading of "financial stability." Financial stability cannot be achieved without a government-wide cost-based accounting system tied to performance within a federal budget ceiling.

The government's accounting systems, in the past and today, have been and are transaction-based. However, their focus has been on controlling the obligation of funds rather than on accounting for cost effectiveness within the federal budget. Because of the long-term nature of major systems and projects, it is necessary to have a management accounting system that will accurately track actual costs against budgetary estimates and provide a credible basis for future cost estimates.

We were astounded to learn, in our study, that senior officials declare that the United States government does not yet have a good accounting system. Without one that is compatible with the budget structure, accurate, and timely, it is impossible to estimate costs accurately; and underestimating the costs of major systems has been a significant contributor to their financial instability. Good cost estimating systems are needed to use the outputs of good accounting systems effectively; and good accounting and budgeting depend, conversely, on good cost estimating. Terrence C. Golden, administrator of the General Services Administration, put the situation into perspective in a keynote speech before the Interagency Committee on Information Resources Management in mid-1986. He said, "The federal government has done a good job in developing the systems to deliver the program itself but probably has spent significantly less time and a lot less money in devoting energy to the financial and management [accounting] systems that are necessary to run the government." He went on to say that "today's financial systems are of the stone age variety. They provide too little information, are typically too late, and are incoherent. Clearly, from my perspective, if we are going to have a simpler, more efficient government in the 1990s, we really have to focus on financial management [accounting] today."

The need for improved financial management accounting systems ranges from the standardization of definitions of cost elements to the accurate tracking of actual cost expenditures to budgets. One government study on productivity[1] stated that "one cannot manage what one cannot measure; and one cannot measure what one cannot define." A process is needed that will define the work to be performed (the dynamic national plan), measure how well it is being accomplished, and manage it within allocated resources. The *Huntsville Times*[2] recently quoted *The Boston Globe* which cited studies by the General Accounting Office, the Congressional Budget Office, and the Congressional Military Reform Caucus that found that from $23 billion to $53 billion could not be accounted for in the Pentagon's budgets for fiscal years 1982 through 1985. The funds were provided to the Pentagon to pay for anticipated inflation that never materialized. Neither the Pentagon nor the House Armed Services Committee was able to determine where the money was spent. Something is wrong with our measuring system if we can't find this much money.

Financial megasystems that accurately track and report costs are being developed in several agencies, such as the Department of Housing and Urban Development,[3] the Department of Transportation as quoted in an OMB document,[4] and NASA.[5] But much still needs to be done to develop accurate cost accounting and credible cost estimating and to tie these measurements to both previous and future projections of expenditures.

The Canadian government has developed an exemplary accounting system that has been proposed as a model for a United States system[6]; several states—California is one—have installed state-wide accounting systems that handle multiple state agencies in an on-line real-time fashion.[7] Government officials, industry suppliers of accounting systems software, and professional accounting and estimating societies all agree that the technology is now available to plan and implement an urgently needed government-wide management accounting system that will track expenditures and report vital information to managers and budget personnel at all levels. The development of this system must be number one on the priority list of megasystems development if the government is to have a tool for the 1990s and beyond to accurately estimate and manage major systems.

In financial systems, as in dynamic national planning, we are forced to go into a higher level of management and policy than the major systems themselves in order to recommend meaningful improvements. The needs of major systems force a discipline on broader issues because the many studies that have been conducted point upward for help in eliminating chronic problems. National goal-setting and planning are needed to implant a more solid base for the start or continuation of major project phases. Likewise, good financial management accounting systems at levels *above* as well as *within* the project are indispensable for accurate cost estimating and cost control. The fact that inadequate planning methods and imperfect accounting systems are problems associated with the entire spectrum of government activities does not dissuade us from proposing sweeping changes if they will cure the ills of major systems acquisition. Undoubtedly many other financial problem areas will be cured at the same time these fundamental improvements are made.

Deep-seated problems in government budgeting, financing, and procurement have effects on our major systems acquisition processes that only the most ingenious and dedicated project managers and administrators can overcome. Hence, we continue to press upward into the realm of government finance to discover if our findings are borne out by others and if there are fundamental changes needed to enhance the estimating and cost tracking of major systems as well as other products, services, and processes in the Federal government.

We interviewed Mr. Fletcher Lutz, executive vice president of the Association of Government Accountants (AGA), to get his reaction to our initial findings about the need for a better accounting system. The AGA, at the time of the interview, was supportive of two bills introduced in Congress, one in the Senate and one in the House of Representatives, for strengthening of controllership in the federal government. The bills attempt to put someone in charge of accounting and financial management for the federal government, and to strengthen these systems. A portion of our enlightening interview with Mr. Lutz[8] follows:

RDS: *One thing we are shocked about and we think maybe the taxpayers will be shocked about when they find out, Fletcher, is that the government does not have an accounting system. Is that correct?*

LUTZ: *That's right. That is correct. There is no general uniform accounting system for the executive branch of the government. That's number one. Number two, every agency has basically done their own system whether it be payroll or personnel or general ledger or cost accounting. Everybody's done their own. I've spent thirty years in this business and at the Department of Energy (I went there to help get a system and get it developed and used properly), it didn't take me too long to find out (two years to be exact) that they didn't really mean it so that's when I decided to retire. I wasn't going to waste my time.*

RDS: *They didn't want it?*

LUTZ: *That's right. They didn't want it. I've done a lot of systems work in government, a lot of it. And as soon as I tell you this you'll understand. In the Civil Aeronautics Board we had a uniform system of accounts for all of the airlines. They were required in the days of regulation to actually follow that system. We were not the only regulatory agency that provided [accounting systems] for their clients that they had jurisdiction [oversight] over. Federal Power, ICC [Interstate Commerce Commission], I mean all those have accounting requirements.*

RDS: *But that's imposing something on industry rather than doing it within the agency.*

LUTZ: *That's correct. That is exactly correct. It's good enough for them but it ain't good enough for us.*

Commentary. It is our view that U.S. taxpayers should be offered the same courtesies as are the stockholders of major corporations. We should each receive a quarterly statement from our government as to how the country's finances are faring. Has the national debt been reduced? If so, by how much? Is the annual deficit being controlled, or is it still escalating? What progress is being made in paying off our foreign debt? All of these questions should be answered on a quantifiable, dollars-and-cents basis. Are our major systems on schedule and within budget? If not, which ones are slipping or overrunning, and what is the new projection for completion? Any citizen who would invest the same amount of money paid in federal taxes in a business or corporation would want an audited financial statement from that company. We, as taxpayers, should require no less from our own government. But to provide this financial statement, the government needs to impose upon itself the same financial discipline it requires of its own contractors. To continue with the interview:

RDS: *Why is it they didn't want to adopt [an accounting system] in the Department of Energy?*

LUTZ: *I think DOE just realized that those things don't come cheap. It takes manpower. It also takes some bucks for equipment and so on. They played around with some of it but they weren't ready to take the whole thing and say, "OK, we're going to really do it," or "Give these people some authority, and we're going to get this thing developed."*

RDS: *Do you think it was manpower and equipment rather than that they just wanted to keep things as they were?*

LUTZ: *I'll be honest with you. I think it was manpower. I think it was in purchasing; in the bucks involved. And I would want to say that from the budget side, and sometimes the administration side, it's much easier if you don't have too much data. You can say more things than you can do. That's why those nice fuzzy budget numbers are what we use. There's some of that [resistance to change] but I think, in all fairness, they still don't have that comptroller's shop in Energy straightened out yet.*

Commentary. It is apparent that federal agencies are motivated by at least three forces that have hindered the development of precise, auditable, and traceable accounting and financial reporting: (1) resistance to change; (2) hesitancy to expose financial history and plans; and (3) desire to build up personnel, resources, and domains of responsibilities. Imposing a rigorous accounting system would place administrators and managers in a position of potentially having to explain and justify the expenditure of all categories of funds in a more organized, systematic, and methodical manner than they are accustomed to doing. A changeover to industry accounting standards would require considerable time and effort. Exposure of past financial actions can sometimes be embarrassing, and exposure of future plans can sometimes be detrimental to an agency, department, or office in its competition for the federal (taxpayer's) dollars and in its flexibility in using these dollars. Exposure of financial status brings to light necessary financial restraint. When confronted with the requirement for more systematic and rigorous accounting, an agency head or office chief has the tendency to request more resources in terms of personnel, facilities, or computer support to activate and run the new accounting system. Huge organizations have huge inertia, and if there is no profit motive or no visible customer (stockholder) to lend support or to complain, business as usual results.

To continue with the interview:

LUTZ: *Now your states are starting to get some good [accounting] systems of their own. True, federal government is a little different but it's really not that much different than some of our big states. Systems can be somewhat similar. You just have to consolidate more.*

RDS: *Are there any states that are outstanding in this area?*

LUTZ: *New York is considered to be quite good and Illinois is considered good. Tennessee is quite good, as are the states of Washington and California. Those are the major large states that seem to be in pretty good shape and are used as examples.*

RDS: *And they do have state-wide accounting systems that take care of everything—a transaction based system?*

LUTZ: *Yes, yes. Virginia is another one. Ed Mazur has done an awful lot to develop that whole situation. He's comptroller of the state of Virginia and a member of our group. He has a very, very capable operation.*

RDS: *Do they have automated systems?*

LUTZ: *Some parts of their system are automated and they are automating more and more. They've been at it a few years, too. It's not like they are just starting. The only reason I don't want to say it is totally automated is that I don't believe it really is. That takes a lot longer.*

Commentary. The fact that some of the larger states are developing highly sophisticated and capable automated accounting systems proves that systematic, precise accounting can be achieved in large government organizations. The economies of scale in software development would make it much cheaper on a proportional basis for the federal government to install automated accounting systems in all of its departments, agencies, and offices than it would for the individual states to do so. This assumes, of course, that all agencies could use the same system. However, if every office contracts for special accounting software development for the year 2000 and beyond, the job could be enormously expensive. Our appeal is that the automated accounting needs of the federal government be established on an overall rather than a piecemeal basis.

RDS: *This bill in Congress (to set up a federal accounting system). We talked to one of the staff members of our Congressman the other day. He pulled it up for us on the computer, and we have a copy of the bill.*

LUTZ: *The Dioguardi [House of Representatives] one?*

RDS: *Yes. Does it propose a major system of accounting or just another organization?*

LUTZ: *It's an organization [only] with direct responsibility to do it. See, that's a big step forward. That is saying, "Hey, we're for this. We want to do it, etc." But we're not doing it. For example in the Roth bill, the President does not want someone telling him how he should be managing, how he should be running his executive branch, and yet it is being run so poorly and nobody knows how much! They don't know how much cash they really have. They don't know how much property they really have, what the value of it is, [even] what the cost of it was which is easier than value. But no, the answer is they [the bills] do not [recommend a government-wide accounting system]. We purposely didn't touch real heavy on it. We didn't want to make everybody in the government mad before they picked up the plan. This plan makes it abundantly clear as to how that kind of a system will work. I must say it is done under the aegis of Reform 88 which is the President's package that tries to get some things done in this area.*

RDS: *Is that to be done by 1988?*

LUTZ: *It's called Reform 88. The theory behind it is that there are a whole lot of projects that are reforming the government finance systems and will be done by 1988, because 1988, of course, is the end of the second four years.*

RDS: *Do you think it is realistic that the government on-line accounting system would be implemented by 1988? If not, how many years would it take?*

LUTZ: *No . . . no way. Even if they started now they wouldn't. No. I would say to do a total Government-wide system will take at least a good twelve to fifteen years. [Note: This indicates that the system would not be ready until after the year 2000] I mean I don't think it can be done . . . I don't mean you can't get parts of it up and running. You can get some things going. I have not seen the uniform system that they have out at Transportation yet. It's fairly new and it's not, by any means, currently adopted by everybody. It won't be. When I say twelve to fifteen years, it means automating—getting it all up, getting it fitting together, getting it reported properly so you can have a total report. Are you familiar with the GAO and Canada project on financial reporting?*

RDS: *I think we saw a copy of it here.*

LUTZ: *You probably did because I have it here. There's an example of some attempt being made [to develop a government-wide financial structure]. In this case Comptroller General Charlie Bowsher and Auditor General Ken Dye from Canada show what is being done in the two countries, what could be done in having one [top level quarterly] financial report.*

RDS: *In other words our government can produce a financial report just like a big company. But it looks kind of funny in there to see the expenditures exceed the income every year!*

LUTZ: *That's right. Exactly, exactly. This is part of it.*

RDS: *This just points it out. It just highlights it.*

LUTZ: *Exactly. As I say, we need some of that. There's another report you should be looking at if you haven't. A pretty good system was developed for GSA [General Services Administration] and there's a lot of variety there, but they are pulling this together and have a financial report for all of the GSA in one report. Other agencies don't do that. The Comptroller General gave them a pat on the back for this (had a little ceremony with it). Price Waterhouse did a lot of the work I believe and I believe Arthur Anderson did some with them. But they have that system up and running and being followed. There's a good example of a rather diverse large operation within the government being able to do this.*

This interview was indeed enlightening. It showed us that a tremendous task is ahead of the government to get itself on track financially. Mr. Lutz was not alone in his insistence on the need for better accounting. Studies by the Grace Commission, the Joint Financial Management Improvement Group, and others verified the need for better government-wide tracking and reporting of costs.

A study done by the Association of Government Accountants[9] entitled *Strengthening Controllership in the Federal Government* emphasized the need for a government-wide accounting and financial management network. This report contains the following six findings:

- *Many departmental and agency accounting systems and systems of internal controls are technologically outdated and relatively inflexible, and provide only minimal assistance to management in their overseeing and monitoring roles.*
- *Key government-wide financial data is not available to the Congress without extensive special analysis and manual effort.*
- *The disclosed error rates of major reporting systems are excessive and costly to the country; such conditions are not restricted to any one department or agency.*

- *The authority and responsibility for data security is split among several officials within the departments and agencies—a difficult situation to correct or improve.*
- *Where some departments and agencies have been fortunate enough to obtain resources to automate portions of their accounting and management information systems,* some of these systems are not integrated or compatible with other automated processes of the organization *[emphasis added].*
- *Since few departments and agencies have integrated or compatible budgeting and accounting systems, few managers are able to effectively manage or provide assurances that Congressional appropriations are being expended as planned.*

The 1983 Grace Commission Report, studies by the General Accounting Office, and work by several privately funded organizations such as the National Academy for Public Administration and the Private Sector Council have shown that a government-wide financial management and accounting system should be a matter of the highest priority. To get it done right, there is only one way it's going to happen. That is, we're going to have to be organized to do it, and we're going to have to have people, provide resources and personnel to develop a system, and make it go.

Over the past 100 years, there have been dozens of major initiatives to improve financial management in government. These include committees, commissions, legislation, and executive orders. Financial management has been studied, exposed, explored, reorganized, analyzed, and changed nearly as much as the government itself. Table 4.1 shows some of the major initiatives affecting financial systems that have occurred since 1894. The biggest change was the decentralization of financial management control from the Treasury Department to the many government agencies,

TABLE 4.1. Chronology of Financial Improvement Initiatives

Date	Title of Legislation or Study	Description/Results
1894	Dockery Act	Eliminated excess offices, provided centralized auditing, instituted preliminary examination of records, and simplified accounts.
1921	Budgeting and Accounting Act of 1921	Established the General Accounting Office. Created the Bureau of the Budget in the Treasury Department.
1937	President's Committee on Administrative Management (Brownlow Committee)	Established the Executive Office of the President (in 1939 the BOB was transferred to the Executive Office of the President from the Treasury Department). Recommended changing the "Comptroller General" to the "Auditor General" which was not approved.

(continued)

TABLE 4.1. *(Continued)*

Date	Title of Legislation or Study	Description/Results
1942–1946	World War II	Decentralization of the Government's financial processes and systems.
1947–1949	First Hoover Commission	Recommended establishing an Accountant General in the Treasury Department which was not approved.
1948	Joint Financial Management Improvement Program (JFMIP)	Combined efforts of GAO, Treasury Department, and OMB in financial management improvement.
1949	Executive Order #10072 Classification Act of 1949	Implemented other Hoover Commission recommendations.
1950	Budget and Accounting Act of 1950	Established JFMIP as a statutory function. Enacted more Hoover Commission recommendations. Institutionalized decentralization.
1953–1955	Second Hoover Commission	Recommended administering and formulating budgets on a cost basis.
1956	Amendment to 1950 Act	Recognized usefulness of and required that budgeting to be done on a cost basis but was never implemented. (Congress did not enact legislation on centralized financial leadership recommended by the Second Hoover Commission).
1953–1959	Rockefeller Committee	Offered other influential financial management initiatives before 1970.
1965	Brooks Act	
1967	President's Commission on Budget Concepts	
1970s	BOB reorganization	BOB reorganized and renamed "Office of Management and Budget."
1970	Legislative Reorganization Act of 1970	Increased Congressional influence on government financial management.
1974	Congressional Budget Impoundment Act of 1974	
1978	Civil Service Reform Act	Proposed initiatives aimed at improving federal financial management.
1978	Inspector General Act	
1981	"Reform 88"	Initiated by President Ronald Reagan to improve federal management.
1982	Federal Managers' Financial Integrity Act	

which started with the Budgeting and Accounting Act of 1921 and continued to manifest itself during World War II. Decentralized accounting was institutionalized with the Budgeting and Accounting Act of 1950. Now, organizations such as the Association of Government Accountants, the Office of Management and Budget, the General Accounting Office, the Treasury Department, and the Joint Financial Management Improvement Program Group are advocating steps that would move toward greater uniformity, standardization, and centralization (rather than decentralization) of accounting, as well as its integration with budgeting.

A Standard U.S. Government General Ledger

The first step toward a more uniform, standardized U.S. government accounting system has been taken in the form of a proposed government standard general ledger.[10] Issued by the Office of Management and Budget to all agencies of the federal government in the fall of 1986, the proposed standardized general ledger has been received favorably by many departments and agencies of the federal government. The goal of OMB is not to establish a single government-wide accounting system, but to establish a single, integrated financial management system for each major agency. Hence, there will be roughly 150 accounting systems, the same number as there are federal agencies and departments. Still, this would be a considerable improvement over the previous and still partially existing situation of multiple financial and accounting systems *within* each agency (according to OMB, there are now 463 accounting systems among the major federal agencies).

The proposed standard general ledger has accounting entry codes for assets, liabilities, equity, budget, revenue, expense, and gains/losses. It breaks each account down to three levels. Each agency will customize the structure, definitions, and breakout to match its own needs and internal accounting structure. Account numbers are suggested, along with "contra" accounts, for each type of entry. Accounting data elements are subdivided into a Fund Classification Structure, a Program Classification Structure, an Organizational Classification Structure, a Resource Classification Structure, and an Other Data Element/Subaccount for reporting purposes. Major systems and projects are to be reported under the Program Classification Structure and broken down into eight data elements ranging from subproject to "budget decision unit." The standard general ledger also has a transaction coding structure and index, and "crosswalks" to standard external reports. The standard general ledger provides a uniform chart of accounts and supporting transactions to standardize federal agency accounting and to support the preparation of standard external reports.

The Association of Government Accountants considers the proposed standard general ledger "definitely a step in the right direction toward greater uniformity in federal financial management and accounting." Some agencies have already adopted portions of the standard general ledger, and others have included it in their five-year plans for financial management improvement. If Fletcher Lutz's predictions are correct, however, it may be ten to fifteen years before a truly universal and uniform federal accounting system could evolve from this effort unless the present and next set of elected officials intervene to expedite solutions.

The standard general ledger represents a uniform accounting *structure*, not a uniform accounting *system*. When new computerized *systems* are adopted to comply with this standard general ledger, some effort is needed to prevent each agency from reinventing the wheel. Customized systems are expensive, while generic multi-use software systems are less expensive because the cost can be distributed among many users. For example, the Peat, Marwick, and Main Company has designed and installed state-wide accounting systems in California and in nine other states. Before engaging software companies to code new systems, a similar effort at standardization of systems should determine and achieve the economies of scale that can be gained in multi-user systems.

To find out how the proposal for a standard general ledger is being received and to determine what steps are being taken or need to be taken toward a federal accounting system, we talked with David Dukes, then executive director of the Joint Financial Management Improvement Program, in his office in Washington. The following are excerpts from this interview[11]:

RDS: *Dave, we have some specific things we wanted to ask you about the Financial Management Improvement Program. For example . . . we're aware that a standard general ledger has been sent out. Have you had any feedback on how that is being received?*

DUKES: *Well, I think for the most part the agencies intend to implement it. Some will have problems that others will not have. Part of the problem is going from the current general ledger set of accounts and resizing that to transfer the data—balances—into the new one. And if the standard general ledger calls for a refinement that the agency might not have, then they've got a problem of analyzing data in their current accounts to make the splitout so it will go into the refined account. I don't have any sense of the difficulties at this point that the agencies may be having. The way we're going to try to deal with this is that the Treasury Department is establishing an implementation working group. They will be meeting with agencies individually and they'll be talking with them about their implementation problems. The Treasury will offer to provide whatever assistance it can. We also expect the GAO to look at what the agencies are implementing, if they are, and how well have they done in implementation.*

Looking at the standard general ledger (360 pages),[10] one can only wonder how much duplicative work is going on in the hundreds of departments, agencies, and offices of the federal government in "implementing" this huge document and in tying the accounts into existing accounting systems. It would appear that, as long as there is going to be a transition to modern, uniform accounting systems throughout the government, more detailed help to the agencies would be needed. It is also not apparent in the standard general ledger that the right categories of cost are to be collected and reported that would enhance the tracking of costs against budgets or the estimation of costs of future projects through comparison with actual history. The new standard general ledger could do a good job of keeping a running record of when and how money is being spent, but there is no proposed mechanism for tracking costs against budget estimates—a vital input needed to achieve financial control. Most companies code the income and expenditures on their major product lines to determine which are profitable. A similar coding for the federal budget could provide visibility as to which systems, projects, initiatives, and services were on track financially, and which were not. Costs to date and run-out cost estimates should be included in an overall federal accounting system to provide an early warning to permit mid-course corrections of programs that are in financial trouble. Continuing with the interview:

DUKES: *The GAO also plans to do financial statement audits somewhat similar to what a CPA firm would do with private industry. They have begun to do that in some areas of government. The General Services Administration, I think, is the largest agency I'm aware of where GAO has done a financial statement audit.*

RDS: *I guess the ideal thing is that everybody will be in a situation where they can have a complete financial audit.*

DUKES: *Eventually, yes. And the other thing that GAO would like to see is to have the Inspectors General turn some of their attention to doing financial statement audits as a way of looking at the integrity of financial operations. GAO has taken the position that, if you focus on the financial statements, you'll get a certain look at the results-oriented aspects of financial reporting. Then you can enforce better discipline back through the system. You just focus on the output and you'll force discipline into your system. There's been an awful lot of focus in federal agencies on the input side but not enough on the output side. GAO is trying to turn that around. I think there's some sympathy for that too. Certainly Treasury believes in that as well with these new Treasury reporting requirements, and I believe that in recent months the OMB has turned to that point of view too.*

Commentary. It is interesting to note that the largest agency on which GAO has done a financial statement audit is the General Services Administration. Why are not all federal agency financial statements audited at least annually if not quarterly? In other parts of this study we found that government is imposing a great deal of auditing on its industrial contractors and suppliers—almost too much, in the estimation of many advocates of procurement streamlining. Then why does the government not perform routine financial statement audits on the big agencies that produce major systems and projects, such as the Department of Defense, NASA, the Department of Energy, and the Department of Transportation? This would appear to be a top priority item once a standard government-wide accounting system is available. To continue with the interview:

RDS: *Dave, in looking at the general ledger we saw some places where funds were transferred from one agency to another. Now is there anybody who really looks at the total top-level picture? Is there going to be a top-level government accounting system that says this is the top-level financial report of the government?*

ALS: *In other words, where does the income show up? Where we pay our taxes—where does that show up?*

DUKES: *The Treasury Department is now producing a report; and, of course, the President's budget also has an analysis that shows the sources of revenues and the application of revenues. Treasury produces a monthly statement—you might just skim through that. [Dukes handed us a copy.] That happens to be the one for September which shows the month of September's activities plus cumulatives for the fiscal year. . .*

RDS: *This is an income statement?*

DUKES: *Well, no, it's an analysis of the financial figures that Treasury accumulates on a receipts and outlays basis.*

RDS: *So it's a monthly financial report. A government-wide monthly financial report.*

DUKES: *Right.*

RDS: *That's great. I'm thinking that there are probably a lot of people that don't know this exists.*

ALS: *Is this not available to the public?*

DUKES: *No, it's available* for *the public. It's not an internal document. It's called the Monthly Treasury Statement. Any citizen can subscribe to Treasury's monthly financial report for $27 per year. Single copies are not available.*

Looking over the Treasury Department's financial report, it is difficult for the layman to find major systems, expenditures, or an easily recognizable income statement or balance sheet. It *does* provide a revealing picture of the deficits being incurred each month. "Outlays" are continually higher than "receipts"—a situation that would be frightening in a corporate quarterly financial statement or stockholder's report. Major systems costs are included, but not specifically identified, in a number of line items under each agency, such as "research and development," "construction," and "procurement." This financial report only tells how much money is spent or received each month. It does not tell: (1) how well we are adhering to budgetary estimates and allocations; (2) how major systems are proceeding; and (3) what reprogramming has been done. According to Dukes, the standard general ledger does not presently include a provision for collecting information on reprogramming actions. (Reprogramming is the shifting of funds from one program to another within an agency's budget.) The federal budget book[12] shows only the actual expenditures for the prior year, not the planned ones. An added column in the budget that shows the funds originally planned for each line item would permit comparisons with actuals to see how well its costs were estimated. This would also help us find out how well we are predicting costs in *other* areas of government. It seems to us that any good budgeting and financial reporting system should be able to reveal our track record. This would provide clues as to what improvements or changes should be made in the future in our estimating procedures. It would also provide valuable information on the mid-course corrections needed to get back on schedule in major programs. Current federal government budgetary and financial reporting systems do not give us a picture of where we have been or where we are going: only where we are now. The standard general ledger, albeit a step in the right direction in providing consistent and relatable information, does little to provide public visibility into the progress of programs. There are models of reports, however, that will provide this visibility if they are conscientiously adopted throughout all agencies that develop or produce major systems. A very good report format is one suggested by the GAO in its report, *Procurement: Selected Acquisition Report: Suggested Approaches for Improvement.*[13]

Acquisition Reports for Major Systems

Although the GAO suggestions were directed primarily to the Department of Defense, all agencies that develop or build major systems or projects could adopt the same standards for financial and progress reporting. Using cost and quantity information from Department of Defense Selected Acquisition Reports for the F-16 fighter from December 1975 through December 1984, the GAO developed graphs and tabular data to demonstrate how financial progress data on large projects could be made easier to understand. The suggested graph formats included a chart showing the original estimate of total program costs versus the "current estimate" of total program costs for the ten-year period 1975 through 1984. The graph was plotted in 1975 dollars to remove the effects of inflation. Even so, the total program cost increased from $4.4 billion in 1975 to $21 billion in 1984. Why the cost increase? Another chart shows one of the principal reasons: quantity increases. The total program, originally scheduled to include 658 aircraft, grew to a quantity of 2,803 aircraft by the end of 1984. That still did not include a full explanation of the cost increases, however. A third chart shows the current estimate of unit cost for each aircraft versus the original unit cost estimate. Cost per aircraft, *exclusive of inflation*, grew by about one million dollars a copy from 1975 to 1984. A fourth chart, a "stacked" vertical bar chart, shows the *reasons* for the cost variances from the inception of the report, and the amount of cost increase caused for each reason. The three largest cost increases were caused by quantity increases (already mentioned), engineering changes, and "other" reasons, which include addition of support costs, schedule changes, underestimates, and "miscellaneous" causes.

The proposed charts are very valuable for project managers and administrators as well as for reporting to the public and to the Congress because the program track record can be easily observed, reasons for cost increases can be identified and dealt with, and future trends can be predicted for the program or for similar programs. The graphs are very simple and easy to understand and, with a good automated accounting system in place, should be easy to prepare using modern computer software. Tabular data showing planned cost, actual cost-to-date, total cost-to-complete, amount over or under the plan, scheduled completion date, and schedule variances can also be easily and quickly extracted. With modern computer and communications technology, it should eventually become simple for each agency to prepare monthly or quarterly financial progress and status reports of each of its major systems and to submit these electronically to a central office in OMB which could publish a consolidated report for submission to the President and the Congress. Other internal agency reports containing information on the funding of the various program phases, status by appropriation, and detailed cost variance analyses could be produced each month or each quarter for internal management use and transmitted to OMB, GAO, or the Treasury Department on request.

For reporting purposes, the Department of Defense identifies major systems as systems in the current five-year defense program that are past the concept formulation phase and are expected to cost more than $75 million for development or $300 million for production.[14] This definition would probably vary among different agencies but could start at this level for initial reporting. As automated systems become available, and as the reporting of top-level financial data on major systems becomes more routine and systematic, the threshold could be decreased to systems costing, say, $5 million for development and $20 million for production. Security constraints may prevent the release of financial data on certain systems, but the President and appropriate committees in Congress should have access to the latest financial, schedule, and status data on all major systems.

Cost Estimating

Let us assume that the current trends will continue and that at some point in the future we will have an excellent government-wide accounting system that not only tracks funds but also provides vital reports to managers on a timely basis. This system will provide valuable historical and actual information for estimating future projects as well as pinpointing areas where immediate corrections must be made. We are assuming that the system will go down to and into the major systems project level; that standardized, low-cost automated financial tracking systems of a generic nature can be provided to many project offices and customized for their use; and that project offices will have the desire and discipline to use them. Going back to the statement, "What we cannot define or measure, we cannot estimate," we see that we now have a basis for defining and measuring where we have been and where we are. Lack of measurement and visibility has been one of the deterrents to good cost estimating. Availability of good accounting systems will help to eliminate one of the major deterrents to good cost estimating: lack of an adequate base for costing. There is no shortage of good advice, methods, techniques, computer programs, books, or educational programs in cost estimating. Over the past several decades, cost estimating and cost analysis have become recognized professions, complete with professional societies, certifications, publications, seminars, conferences, and academic courses.

One of the most elusive bits of knowledge to the human race has been how to estimate accurately the costs of large projects and how to manage these projects so that they stay within their originally estimated cost targets. Historian Edward Gibbon wrote in *The Decline and Fall of the Roman Empire*: "When it was evident that costs would be more than double the original estimate, the officers of the revenue began to murmur." Commenting on the costs of military weapons and logistics to the British, Benjamin Franklin wrote to Joseph Priestley in 1775: "Britain, at the expense of three millions, has killed 150 Yankees in this campaign, which is $20,000 a head." (Twenty thousand dollars was equivalent to about $300,000 in present-day dollars.) Franklin went on to extol the fact that Britain's gross national product would not be enough to annihilate the colonial armies. As weapons and other major systems grew in cost and complexity and were produced in larger quantities, new tools were developed. Learning curves, estimating relationships, and critical path scheduling methods were conceived as early as the 1930s and later formalized and used in government contracting for large systems. World War II saw the emergence of military systems cost analysis. Pioneering work was done by Ed Paxon and David Novick of the Rand Corporation in developing analysis techniques and identifying the cost elements for life-cycle costs of major systems in the early 1950s. It was during the late 1950s that the Navy developed the Project Evaluation and Review Technique (PERT) which was closely followed by a PERT/Cost Technique that permitted the estimation of costs over time of a project scheduled on a critical path basis.

Secretary of Defense Robert McNamara approved a Program Planning and Budgeting System (PPBS) in 1961, which was followed in the late 1960s and early 1970s with formalized contractor cost reporting systems. An updated *Armed Services Procurement Regulation Manual for Contract Pricing* was published in October 1965, and the Rand Corporation published its classic summary of contemporary cost analysis techniques in *Cost Considerations in Systems Analysis* by Gene H. Fisher in December 1970.[15] Also in the early 1970s, a new concept called "zero-based budgeting" was introduced. This technique was deemed necessary at the time because federal agency budgets continued to grow from each previous year rather than expand or contract to accommodate fluctuating needs. Although the concept didn't last long (i.e., agencies still base their next year's budgets on growth from previous year's budgets), the budget initiatives gave greater impetus to improved cost estimating, scheduling, and program/project definition. Life-cycle costing, as a formalized way of estimating and managing the lifetime costs of major systems and projects, was introduced in the early 1970s, and two books (*Design and Manage to Life Cycle Costs* by Benjamin S. Blanchard, 1978,[16] *Life Cycle Costing: A Better Method of Government Procurement* by M. Robert Seldon in 1979[17]) documented the methods and techniques of life cycle costing.

Cost modeling had started as early as the 1950s with models such as "PROM" developed by the Rand Corporation for the Air Force. Early versions of these models relied principally on hand computations, but soon, in the 1960s, a proliferation of cost models became available on mainframe computers. "Design-to-cost" as an overall philosophy and goal was established in the 1970s, but agencies are still trying to determine how to implement an effective design-to-cost procedure in an environment where systems commitments are frequently based on concepts that are still in the laboratory (more about this later). Also, in the 1970s the minicomputer became available and was used increasingly for cost analysis, cost estimating, financial reporting, and cost management. The 1980s were definitely the years of the microcomputer in cost estimating as in other fields. A proliferation of computers, estimating software, scheduling programs, and customized models became available to permit cost analysts to do detailed estimates and analysis. A number of estimating books were published by your present authors and by others in the 1980s, adding to the already large volume of published literature on the subject of cost estimating. The 1990s promise to be the years when new technologies such as artificial intelligence and the use of computer-based expert systems will come into their own in cost estimating and cost analysis. Automated estimate generation from computer-aided drawings, improved graphics, and multidimensional data bases will be possible on the higher speed, higher capacity personal and mini computers already available.

The National Estimating Society, formed in San Diego in 1966, now offers a certification examination for Certified Professional Estimators, and the Institute of Cost Analysis, located in Alexandria, Virginia, awards a "Certified Cost Analyst" designation to those who pass a rigorous examination. These and other organizations, such as the American Association of Cost Engineers and the International Society of Parametric Analysts, hold conferences, give seminars, offer training, and produce publications that tout the latest estimating techniques and tools.

So the problem in preparing good cost estimates for large government projects is not the result of a lack of methods, tools, techniques, or even qualified people. Why, then, are there still persistent reports in the press, in government auditing reports, and in Congress about chronic underestimation and subsequent overruns in large projects? The General Accounting Office (GAO/NSIAD-87-128) recently reviewed nineteen major defense acquisition programs scheduled to reach either full-scale development or full-rate production by fiscal year 1988 or 1989. Nine of the nineteen programs had increased baseline cost estimates, and twelve had experienced schedule slippages. The average delay of those that slipped was seventeen months, and in most cases the delay resulted in increased costs. The Advanced Medium Range Air-to-Air Missile experienced significant cost growth and schedule delays. Its estimated cost has more than doubled—from about $3.4 billion for about 20,000 missiles to $8.2 billion for 24,335 missiles. The reasons cited by GAO for the understated cost estimates included:

1. Competition for program funding and pressure to justify the program over new or existing programs.
2. Contractor competition for the missile design and development contracts.
3. Frequent project manager turnover.

All three of these reasons—and suggested ways of counteracting these effects—are addressed in this and other chapters. Attaining financial stability will require that discipline be exercised in making more realistic commitments for programs at their outset as well as the dedicated use of tools, techniques, and skills that are already available.

Given that improved accounting and financial systems will provide estimators with sound and reliable data, what steps need to be taken to assure that accurate, realistic, and achievable program cost estimates are (1) consistently *produced* and (2) consistently *used* in planning and carrying out programs? The answer lies in the method of definition and estimation of programs on the one hand, and in disciplined adherence to program estimates and plans on the other.

By far the most accurate, credible, and achievable cost estimates have come from the use of detailed engineering-based estimates of projects that have been thoroughly conceived, analyzed, designed, and planned before the estimate is prepared. In these estimates, exact material quantities as well as predicted scrap and waste are estimated from plans, drawings, and specifications of the tangible elements of the project. Intangible elements are estimated on a task/skill basis. Work is subdivided into enough detail to permit the estimator to determine how many minutes, hours, or days it will take each person or organizational entity (crew, office, shop, or team) to do the work. Armed with the person-power needed, the labor rates for each skill or team, and productivity or efficiency factors based on past and projected performance, the estimator can develop precise labor cost estimates. Material and equipment cost estimates based on quantities and projected prices are combined with labor cost estimates and other costs, such as travel, transportation, subcontracts, and services, to produce the total direct costs of the job. To this is added overhead, general and administrative expenses, and fee or profit if the activity or project is performed by industry. This type of estimating has been used successfully in many construction and industrial programs that have outstanding records of cost containment, frugality, and efficiency. What, then, is the problem? Why haven't we been able to make accurate estimates on more programs or to stay within these estimates once they were established? The problem lies in the *definition* of the projects, both initially and as they are carried out!

To get a good initial cost estimate, a very detailed definition of the project, its design, and even its operation is required. Achieving this detailed definition for a megaproject usually requires much work in itself. In construction, this definition phase is done by an architect-engineering firm. The definition phase involves detailed engineering analysis, design, drawings, interfacing with the customer or user, the building of models or mock-ups, site surveys and legal arrangements, detailed specification preparation, and even initial contract drafting. For high-technology weapons systems, communications systems, or information systems, the initial detailed definition may even include the construction and testing of laboratory or full-scale prototypes which, in themselves, can be very expensive.

In the meantime, far ahead of this detailed definition period, program advocates are trying to get the *entire project* included as a line item in the federal budget. Companies who may participate, anticipating a possible large profit, are already putting pressure on Congress to include the item in the budget. Since industry is the major supplier of government projects, it most probably was the source of the idea in the first place. The question always arises, then, at this early stage, of "How much will it cost?" with very little definition of what "*it*" really is! Government analysts and their industry counterparts are frantically searching for what is then a comparatively sketchy definition of the project on which to base a cost estimate. Parametric estimates, based on preliminary performance parameters and past history rather than a detailed design and program plan, are used in these early program phases. They are based on system weight or performance, or combinations of these and other factors. Parametric estimates have proven to be useful to determine the cost of a typical or average program based on very early requirements only; but since the specific design is a large determinant of cost, parametric estimates can miss specific program targets by a wide margin. Program advocacy and inadequate definition in early stages have often caused these parametric estimates to be unrealistically low.

Since we can only estimate what we can define, then, the solution of the initial program cost estimating problem is to plan, estimate, and firmly approve programs only on a phase-by-phase basis rather than on a total program basis. Since earlier program phases are less costly than later program phases, this incremental approval route will result in far less wasted effort than sporadic approval, disapproval, or stretching out of entire programs. The required analysis can be funded without committing to the full design effort; the full design effort can be funded without committing to the prototype testing; and prototype testing can be funded without committing to full production. Thus, a stepwise progression of system definition will permit informed decisions, accurate estimates of the next phases, comparison between competing alternatives, and minimum life-cycle investment in potentially unsuccessful system candidates.

The incremental approach must be accompanied by a preliminary estimate for the project as a whole. This preliminary estimate is needed to determine if the project fits within overall budget allowances for the agency, and it is updated as the project proceeds. Cost-benefit analyses and cost-effectiveness analyses of alternatives are absolutely essential at the beginning of the program and at each significant point where project continuation decisions are made. It is in making this preliminary estimate that the parametric type of estimate can be most effectively used. But more detailed estimates must be developed as definition proceeds into further depth. As shown in Chapter 2, projects which have experienced successful cost containment have been those in which design and configuration were frozen early and in which "gold plating" and "tinkering" were discouraged by hard-nosed project managers and others who were brave enough to say "no" to the innovators. Cost estimates cannot be expected to remain valid if the pedigree of the project changes. In some instances, projects have retained their name only, and their definition has changed so dramatically that the "before" and "after" configurations have little resemblance. Only when we are willing to discipline ourselves to true phased procurement with a deepening baseline definition and more detailed estimating at each step will we be able to demonstrate consistently that government projects can be estimated accurately and carried out within budgeted allocations.

Let us assume that we can develop realistic, credible, and accurate estimates on *all* of our big federal projects by using phased estimating, detailed definition, and disciplined restraint to costly changes. The next major step is to assure that these estimates are *used* in the budgeting process. This, again, will require much self-discipline in the Congress and in the executive agencies. The practice of quoting low to win the project and then charging actual costs later must eventually become extinct if we are to achieve full project financial stability. The General Accounting Office has reported[18] that the independent government estimates made *within* departments and agencies are often not used. Independent cost estimates are sometimes higher than program office estimates and are often unused, not because they are inaccurate, but because program advocates fear that the quotation of higher cost numbers will cause termination or postponement of the project.

A more consistent use of independent cost estimates, even though they may be higher than program advocates desire, can be gained through convincing the Congress, the public, and the industrial community of the long-term benefits of realistic cost estimating. A hungry industry can be expected to be optimistic or even overly optimistic when it comes to estimating programs, but it is counterproductive for the government to compound the problem by ignoring independent estimates of its own internal experts. Education of the public, the Congress, and the industrial sector on the importance of preparing solidly based independent estimates and *using* them are essential parts of any plan for long-term improvements in major systems financial stability. Along with the commitment to *use* realistic independent estimates must come a renewed commitment to keep the estimating groups independent and free of pressures to "shave" cost estimates to help the programs through their initial budget approval cycles.

The General Accounting Office, in a report titled *Theory and Practice of Cost Estimating for Major Acquisitions,* dated July 24, 1982,[19] set out nine fundamental and critical criteria for an effective cost estimating process:

1. Clear identification of task,
2. Broad participation in preparing estimates,
3. Availability of valid data,
4. Standardized structure for estimates,
5. Provision for program uncertainties,
6. Recognition of inflation,
7. Recognition of excluded costs,
8. Independent review of estimates, and
9. Revision of estimates when program changes occur.

All nine points are important, but the first, *clear identification of the task,* has more bearing on achieving a good estimate than any of the others. It must seem incredible to the average citizen to hear that some government agencies proceed to budget for programs they have not yet defined—but this is all too true. According to one highly regarded cost analyst who is inside one of the largest government agencies, we have continued to commit ourselves to major systems based on technologies that have not even been demonstrated in the laboratories. Absence of clear definition and excessively optimistic projections of performance, cost, and schedules are undoubtedly the major reasons for underestimates and the resulting overruns and unanticipated cost growth. We asked this same analyst about the "design-to-cost" policy that was implemented from the top down several years ago to hold costs within projected budgets. The analyst told us that in 1986 the agency headquarters sent an "expert" on design-to-cost to the field installation to determine how to implement the policy. The analyst never heard again from the headquarters expert and told us that "design-to-cost" is a dead issue. It is not being implemented in her agency.

There is obviously an alternative to the helter-skelter rush to approve whole programs based on unknown technology—the phased acquisition approach that permits detailed study, research, development, and testing in phases to permit credible budget estimates to be developed as definition proceeds, the principal thrust of what has been said in Chapter 3.

Broad participation in preparing estimates means that program advocates and nonadvocates alike must participate in order to maintain some degree of realism in cost, schedule, and performance projections. It might even be wise to let program peer groups produce independent estimates of programs or program phases—if, indeed, one is required while the technology is still in its embryonic stage. An adversarial peer-review would be very healthy in surfacing potential problems—places where cost estimates are not realistic—and to identify overoptimistic performance claims. Availability of valid data depends upon the rigorous tracking and reporting of financial data on existing ongoing programs—an action that will become possible when auditable government-wide accounting procedures are in place. A standardized structure for estimates as well as for accounting systems will permit easier identification of pockets of optimism, absence of credibility, and inadvertent or careless omissions.

One of the most frequently blamed culprits in underestimation is uncertainty in the program. The most effective provision for program uncertainty is to reduce or eliminate uncertainties through the application of rigorous and effective systems engineering, analysis, simulation, testing, redesign, and quality control techniques. Certainly, there will be a need to provide contingency resources for totally unforeseen natural disasters, strikes, and fluctuations in the economy. But high-technology tools and methods are available to develop high-technology systems; and uncertainties can be sequentially weeded out in a phased engineering approach.

Recognition of inflation is a must in major programs. The all-too-common government practice of quoting program costs based on the dollars of several years ago is both misleading and disastrous to the program. Program costs should be both estimated and budgeted in "real-year" dollars, with inflation effects constantly updated and included in the budget. Estimation in real-year dollars will help avoid the surprises often encountered when conversion is made from a past year's baseline estimate to the amount that must actually be paid out by the federal government for the system or project.

Independent review of estimates should mean review by non-program advocates, perhaps even by adversaries. After all, if we are to combat pervasive and symptomatic over-optimism, we must counterbalance it with some conservatism (maybe even over-conservatism) and realism. Since competitive optimism has been identified as a major cause of under-estimates, program advocates should be forced to fight it out in the trenches with program adversaries before Congress is asked to commit valuable and scarce taxpayer dollars to new or continuing programs. Estimates must be dynamic. They must be continually updated to eliminate surprises. This can only be done through application of the strictest discipline in accounting, estimating, and financial reporting. Design, schedule, or performance changes that affect cost should be accompanied by their cost impact, and cost increases should be surfaced all the way from the engineering change board to Congress if necessary to keep costs under control.

Most agencies have detailed guidelines for cost estimating and cost analysis, many of which have been prepared at considerable taxpayer expense. The Institute of Cost Analysis, in Volume II, No. 4, of its newsletter,[20] listed no less than seventy-nine planned and in-process contracted studies on new cost techniques, data base development, specific models, acquisition economics, and operating and support costs for the Department of Defense alone. These studies have been completed, are in process, or have been planned by the Office of the Secretary of Defense (OSD), the Army, Navy, and Air Force. All three services are studying software cost analysis and estimation methods. Hundreds of thousands of dollars are being spent annually on carrying out and improving DOD's cost estimation capability. Are improvements really being made in these processes, or are the studies merely re-documenting old problems, reinventing solutions, or duplicating other parallel work? To determine what type of work has been done and what type of documentation typically exists, we reviewed a two-volume, 1,179-page cost estimating handbook prepared for the Air Force Systems Command.[21] The two-volume book represents one of six multiple-book volumes of methods, techniques, and data required to estimate the Air Force's major aeronautical, armament, electronic, missile, and space systems. The three-inch-thick first volume of the two-volume set describes how the estimating process now works in the Air Force as well as in many other agencies of the government. Government-wide publication of these estimating methods and standards, along with guidelines that attack the root causes of underestimates, such as inadequate definition and program advocacy would be beneficial to all agencies in stepping up to the job of improving the government's cost estimating capability.

In GAO reports on cost estimating techniques we found a desire (and an agreement by responding agencies) to use "independently developed" cost estimates. We found very little in the way of a good definition of an independent estimate. In some instances, *independent* merely meant that the estimate was prepared outside of the program office. In other instances, the independent estimating office was in the parent agency or department of that which was advocating the program. For example, the Office of the Secretary of Defense has a Cost Analysis Improvement Group that prepares independent estimates of programs proposed by the services. The question that arises is, "How far must the 'independent' estimating office be removed from the proposing office or agency so as not to be influenced (in submitting lower costs) by the advocating agency or office?" The answer lies not in the organizational location of the estimating office but in the motivation and commitment to cost realism engendered from the top down in the procuring organization. Cost realism is only possible if the Congress, the executive branch, and the public recognize the devastating effects of over-optimism and refuse to permit the start of programs or program phases that are not fully defined and accurately estimated. The criterion for the start or continuation of major programs, in a limited budget environment, must be much more restrictive than they are now. One of these criteria is the preparation *and* use of credible and realistic independent estimates prepared by the government itself.

Budgeting

The ability to budget realistically is a much sought-after objective. In the Defense Acquisition Improvement Program, also known as the "Carlucci Initiatives,"[22] over half of the thirty-three initiatives advanced by then Deputy Director of Defense Frank Carlucci in 1981 for the Department of Defense, were directly or indirectly related to budgeting. However, only those initiatives associated with reorganization and acquisition decision-making have been fully implemented. The recommendations related to budgeting have not been fully implemented to the satisfaction of GAO for various reasons. Some were not practically achievable without full cooperation of the Congress. Others required management philosophies of cost realism, detailed definition, phased procurement, and multiyear procurement that were not fully supported throughout the Department of Defense, in OMB, or in Congress.

Six of the initiatives targeted to be addressed by the Department of Defense were *directly* related to budgeting. The principal problem that budgeters have had in major systems is that the budget cycle requires inputs way ahead of even the earliest phases of definition. One feasible solution to this problem is to provide major system funding ceilings without allocating costs to specific programs. Then, specific programs and program phases can be prioritized and selected *as they are defined in detail*. In a "budget-to-most-likely-cost" initiative, the services are required to submit two estimates for programs in the formative stages: a program office estimate and an independent estimate. Since 1984, the DOD authorization acts have required the Secretary of Defense to obtain and consider independent cost estimates before approving the full-scale development or production and deployment of major acquisitions. Ostensibly, the independent estimate will be an accuracy and credibility check on the program office estimate because it is not prepared or approved by program office advocates. Unfortunately, it is usually lack of detailed definition of the product or activity rather than solely program advocacy that causes the estimates to be too low. The problem has been that independent estimates are based on the same inadequate definition as those of the program office, since detailed definition has not been done by either group this early in the program. Hence, it is better to provide overall major system budgetary ceilings than to preselect major systems—and, as a result, increase program office and contractor hopes—based on inadequately supported and unrealistically synthesized program office and independent estimates. This will require steadfast patience on the part of the system advocates, but it will result in more credible budgetary estimates for all programs in the long run.

So far, no one in the cost estimating or budgeting community has devised an infallible method of quantifying the cost of "risk,"—although many have tried. Invariably, the costing of unknowns has resulted in an unknown cost. In an initiative called "Budgeting for Technological Risk," the DOD is attempting to quantify the contingency funds needed due to uncertainties inherent in developing major weapons programs. Regrettably, it is virtually impossible to exactly identify required contingencies when a project is not fully designed. For example, if definition is 85 percent complete, the remaining 15 percent of the design or definition process may result in much more than 15 percent added cost. Large cost drivers may not be defined until the very last touches are put on the specifications and their resulting designs. Hence, "risk," which is partially related to the undefined portion of the project, is hard to quantify. A way to develop more credible cost estimates in this context is to reduce or eliminate risk by more fully defining the product or project prior to entry into full scale development and by completing the design, development, and testing of the product before committing to full scale production.

The policy of "baselining" (establishing or "freezing" a set of requirements, configuration, or a design) is a good idea in general, but its definition and effects must be fully understood before the policy can really be implemented. Large projects and systems go through a metamorphosis as they proceed from idea through concept, design, hardware, and operation. They begin with almost nothing defined and should end with everything defined. As they proceed from germination at the beginning of the cycle to fruit-bearing at the end, the baseline changes from that of a list of requirements to that of a manifested operational system, facility, or process. Greater focus and definition is possible as the final product matures to a usable item. The important objective is baseline *continuity* or *consistency* throughout the time period of creating a new project. As in any birth, the general configuration is established at the time the seed is planted, and then it grows in detail as development proceeds. Thus the baseline evolves. "Freezing" a baseline at any point in the project simply means that, for maximum economy, we should not make major changes or additions to the general configuration after the project has reached predetermined milestones. Thus, the baseline is dynamic in that it increases in depth or degree of definition as the program proceeds. DOD has recognized that baselining is required to achieve more realistic budgets but has not enforced its implementation in a stepwise basis in every major acquisition. GAO states[19] that DOD's baselining initiative results "have been disappointing because of less than complete implementation." *Requirements* must be baselined before going into preliminary design; *preliminary design* must be baselined before going into final design; *final design* must be baselined before going into development and testing; and so on. A minor disruption of this cycle (changing requirements just before going into production, for example) can have calamitous effects on program cost. Also, sequential baselining of requirements, design, and final configuration must go hand-in-hand with the baselining of the budget. As definition is deepened, greater budget detail must be made available, and the budget itself must be baselined.

"Design-to-cost" has been a buzzword for many years in the systems acquisition community. DOD's initiatives include a policy to "provide more appropriate design-to-cost goals." The purpose of this initiative is to better control weapon system costs by "providing contractual incentives to industry that more closely associate design-to-cost (DTC) goals with actual costs incurred."[19] The General Accounting Office states, "DTC incentives are not being monitored by the Office of the Secretary of Defense (OSD), and responses to our questionnaire indicate that these incentives are not widely used." We believe that the reason that this technique is not widely used is that there is not a full understanding of the process. The concept of design-to-cost is based on the premise that budgetary cost limitations or targets imposed on a program can be reflected back into its design, quantity, quality, schedule, or specifications. Workable design-to-cost principles require that at least one detailed design and corresponding cost estimate be available. One or more program options are then reviewed, adjusted, or merged to produce a program that fits available resources. Meticulous traceability of costs and their association with performance, schedules, and quantities is needed for design-to-cost goals to work. Thus, the same conclusions keep emerging: (1) we must define, baseline, and estimate our projects thoroughly at each step, and (2) we must have the proper accounting and financial reporting tools available and in use to establish budget credibility, accounting accuracy, and financial stability.

The Department of Defense's initiatives also deal with the integration of the acquisition decisionmaking process with the budgeting process. Simply stated, this is an effort to "live within our means." Both the DOD and the GAO are attempting to measure the effectiveness of this initiative by determining the amount of reductions in new starts, but this still may not result in an affordable overall program. The criteria for success should be whether we can stay within our budget, not whether we can successfully limit new program starts. Limitation of new starts may be an outcome, but it may not be the appropriate criterion for budget control. According to the DOD Inspector General,[19] the DOD has "not consistently followed procedures designed to ensure that sufficient resources are committed to carry out major acquisitions." In the past, many programs have been inadequately funded because more were started than could be accommodated in the budget. The most detrimental form of financial instability is lack of funds. And if we continue to start more programs than we can afford, we will continue to encounter budget-busting financial instability.

Other DOD initiatives indirectly relating to budgeting include preplanned product improvements, multiyear procurement, increasing program stability, encouraging capital investment to enhance productivity, establishing economic production rates, reducing administrative costs and time to procure items, reducing governmental legislation and regulations related to acquisition, forecasting business base conditions at major defense plants, and increasing competition. Many of these initiatives deal with the symptoms rather than the root causes of financial stability, and some are not realistically achievable without sweeping changes in management philosophy. As mentioned earlier, and as will be discussed several more times throughout this book, the problems in major systems acquisition cannot all be solved *within* an agency or department because of the integral role of many other players in and out of government (the Congress, industry, and other agencies). This is why broad changes in management philosophy, not organizations, will be needed. The lessons learned in the Department of Defense can be applied equally to other agencies that plan, acquire, and operate major systems and projects, because these other agencies encounter many of the same problems and pressures.

Since the Congress authorizes and appropriates funds for the development of major systems, it plays a key role in the budgeting process (see Chapter 7). It is important for each member of Congress to realize fully the effect of *competitive optimism* on the budgeting process, and to resist those who want to put unrealistically low estimates in the budget to help get a program approved. It is important that Congressional subcommittees, committees, and individual senators and representatives be provided with highly credible detailed estimates prepared by truly independent cost estimators who are nonprogram advocates before voting funds for major programs. *The only way that we see this to be realistically achievable is for each member of Congress to receive steady pressure from his or her constituents to resist unrealistically conceived, inadequately planned, or overzealously promoted programs based on yet-to-be-developed technologies or inadequately engineered and tested concepts.* This will require steady pressure from an informed electorate. It will require a public that is concerned and knowledgeable about the problems in buying big systems, their causes, and their potential solutions. It will require citizens who actively promote (even demand) objective, vigorous, detailed planning, scheduling, definition, and estimating of new major programs. And it will require elected officials who will respond to these cries for more realistic budgeting and include the proper budgetary and estimating cross-checks in the budget formulation process. In our view, this will require more rather than less financial micromanagement by the Congress unless the executive branch agencies can eventually build up their reputation for accurate estimation and precise cost management.

Just as important as realistic initial budgeting is the need to enforce discipline to stay within budgets. The budget structure, as well as the accounting structure, should be performance-oriented to permit the funding of objectives or functions in addition to the funding of people or organizations. In other words, financial structures should be output-oriented rather than input-oriented. Budgets must be derived with the knowledge that they are not limitless sources of funds. On the contrary, managers need to know when they approve an estimate of the program and commit to perform it that they will be held responsible for conducting it within the requested funds. Knowledge that the available funds will be limited to the forecast ceiling amount is an effective deterrent to underestimation at the program's major decision points.

Financial Stability Requires Program Stability

An important DOD initiative which, in fact, needs to be a government-wide initiative is to *increase program stability.* Because the Congress has the role of approving funding for systems and projects (see Chapter 7), it is not possible for the Department of Defense (or any other agency) alone to establish program stability. The General Accounting Office reports this initiative as "not fully implemented" and that "DOD has made limited progress in stabilizing programs despite unprecedented increases in defense spending since 1980."[19] GAO also states, "DOD has reported essentially no progress in stabilizing major weapon programs." DOD has been caught in a "vicious cycle" of cost increases and program instabilities as described by GAO as follows[19]:

DOD has attempted to improve program stability through some of the other initiatives included in the acquisition improvement program with limited success. Specific objec-

tives included budgeting more realistically by using less optimistic assumptions, reducing the number of new major programs, and canceling low priority major acquisition programs. Overly optimistic budgets build in program instability because, as higher program costs become apparent and difficulties arise requiring greater than budgeted funds, DOD tends to stretch out programs, that is, acquire weapons over a longer period than planned, to stay within budgeted funds. This leads to higher program costs when the weapons' procurement rates are reduced below economic production levels.

This vicious cycle of overoptimism, program overruns, cancellations or stretchouts, and ultimate cost increases can only be broken by adopting far-reaching management improvements such as government-wide long-range major acquisition planning; industrial-type accounting and estimating systems; project management continuity; elimination of counterproductive procurement practices; and a planning and approval process that will result in an unswerving government-wide commitment to successfully and economically complete each major project. A phased system/project approval cycle that methodically selects only those projects that can be started or continued under an overall budget ceiling has the potential for providing the type of financial stability that will consistently produce successful results. By continuing to emulate successful programs, and to eliminate the root causes rather than just the symptoms of financial instability, we can provide efficiency, economy, and effectiveness in our major acquisitions. A new accounting system tied to the budget will reveal, on a continuous basis, the programs that are being well managed. Financial reporting systems will reveal the self-destructive results of underestimation and its devastating effect on financial stability. Holding the overall financial posture of major systems acquisitions continually before us will provide the sustained attention needed to correct problems in real time for ongoing projects and systems and help avoid underestimations in the future. With accurate, credible, real-time financial systems in place, discipline can be brought into the system acquisition process by allocating our finite resources to specific jobs or objectives, and then holding managers responsible to stay within budgets.

Endnotes

1. D. Scott Sink, Sandra J. DeVries, and Thomas C. Tattle: "An In-Depth Study and Review of State-of-the-Art and Practice Productivity Measurements," *IE Management News,* AIIE, Norcross, GA, Winter 1985, Vol. XIX, No. 2.
2. Fred Kaplan: "Pentagon is Missing Up to $32 Billion, Study Says," *Huntsville Times,* Huntsville, AL, September 3, 1986.
3. George L. Farnsworth: "HUD Implements Commercial Accounting System," *Government Computer News,* Silver Spring, MD, November 7, 1986.

4. Office of Management and Budget: *Joint Financial Management Improvement Program (JFMIP)—Annual Report 85,* Office of Management and Budget, Washington, DC, 1986.
5. Lew S. Lauria: *NASA Financial Management Systems,* National Aeronautics and Space Administration, Washington, DC, December 86.
6. General Accounting Office and Auditor of Canada: *Federal Government Reporting Study—Summary Report,* General Accounting Office, Washington, DC, 1984, AFMD-86-30.
7. Interview with David Dukes, Director: Financial Management Improvement Program, Washington, DC, December 8, 1986.
8. Interview with Fletcher Lutz, Director: Association of Government Accountants, Washington, DC, August 20, 1986.
9. Association of Government Accountants: *Strengthening Controllership in the Federal Government,* Association of Government Accountants, Washington, DC, May 1985.
10. Office of Management and Budget: *U.S. Government Standard General Ledger,* Office of Management and Budget, Washington, DC, August 1986.
11. Interview with David Dukes, Director: Financial Management Improvement Program, Washington, DC, December 8, 1986.
12. Office of Management and Budget: *Budget of the United States Government FY87, 4 vol),* Office of Management and Budget, Washington, DC, 1986.
13. General Accounting Office: *Procurement Selection Acquisition Report: Suggested Approaches for Improvement,* General Accounting Office, Washington, DC, July 1986, NSIAD-86-118.
14. Department of Defense: *Preparation and Review of Selected Acquisition Reports,* Department of Defense, Washington, DC, May 1980, 7000.3-G.
15. Gene A. Fisher: *Cost Considerations in Systems Analysis,* Rand Corporation, Los Angeles, 1970.
16. Benjamin S. Blanchard: *Design and Manage to Life Cycle Costs,* M/A Press, Portland OR, 1978.
17. M. Robert Seldon: *Life Cycle Costing: A Better Method of Government Procurement,* Western Press, Boulder, CO, 1979.
18. Office of Management and Budget: *Department of Defense Needs to Provide More Credible Weapon Systems Cost Estimates to Congress,* Office of Management and Budget, Washington, DC, May 1984, NSIAD-84-70.
19. General Accounting Office: *Theory and Practice of Cost Estimating for Major Acquisitions,* General Accounting Office, Washington, DC, July 1972, B-163058.
20. Institute of Cost Analysis: *Newsletter of the Institute,* Institute of Cost Analysis, Alexandria, VA, August 1986, Vol. II, No. 4.
21. Analytic Sciences Corporation: *AFSC Cost Estimating Handbook Series,* Analytic Sciences Corporation, Reading, MA, 1985.
22. General Accounting Office: *Status of the Defense Acquistion Improvement Program's 33 Initiatives,* General Accounting Office, Washington, DC, September 1986, NSIAD-86-178BR.

CHAPTER FIVE

The Project Manager

The person who has the greatest influence on the outcome of a project is the project manager. The manager must, using the best available management and engineering techniques, reduce uncertainty at the beginning and carry out the project while adhering to initial cost, schedule, and performance goals. Lieutenant General Hans Driessnack, speaking before the first annual symposium of the Institute of Cost Analysis,[1] said: "Improvement programs of the past have had limited success primarily because insufficient attention has been paid to the *people who have to implement the improvements* [emphasis added]." The key people who have to implement improvements in the government acquisition process are *the project managers and their teams*! Project managers for super projects must be super people! They must possess a number of important intangible qualities such as dedication, persistence, patience, vision, drive, determination, and enthusiasm. In addition, they must have in-depth knowledge and experience in project management techniques and in the engineering or technical fields needed to carry out the project.

Successful project managers usually have advanced from positions in design, construction, or field engineering through the engineering and management of increasingly larger projects in either industry or government. They believe in their projects, and they are willing to put forth extra effort to see that their projects are completed on time, economically, and successfully. Within these basic traits and characteristics, there are many acceptable management styles or approaches, but the successful project manager usually develops in-depth knowledge and enthusiasm for his or her project. One such project manager is Mr. Dave Shilling, construction project manager for a portion of the Washington D.C. Metrorail project. The following is a account of impressions gained during a visit to his field office and the actual work site:

In late August 1986 we drove to the mobile office trailer of Metro Construction where Dave had arranged a tour of their new facility in Wheaton,

Maryland. [*Lesson 1: Effective Project Managers Are Open and Willing to Talk About Their Projects.*]

During the preliminary interview he described the construction methods being used on the Metro project. He showed us samples of the materials and interpreted the comprehensive schedule charts that wrapped the room. He spent a considerable amount of time explaining the New Austrian Tunnel Method (NATM) which was used in this segment of the Metro. The low bidder, Ilbau America (based in Austria), came in at $30 million below the engineering estimate and proposed a value engineering change that further reduced the price to some $50 million. The Metro had had water problems because of the infiltration of acid from iron sulphides in the local soils, which chews away the reinforcing bars. In the NATM, after each blast, the ground movement was stabilized by placement of a quick-curing support material similar to shotcrete. A steel reinforced concrete liner was not necessary. At the Wheaton Station the shotcrete-like material was further reinforced by welded wire fabric. When excavation was finished, waterproofing was accomplished by placing a felt-like drainage fabric about 1/4 inch thick over the shotcrete. This was then covered by a PVC [polyvinyl chloride] layer held in place by rubber or plastic capped spikes; seams were heat-welded. Dave was excited about this construction approach, which really wasn't new (just new to the United States), and said that it saved both time and money for this project. [*Lesson 2: Effective Project Managers Know Both the Technical and Business Aspects of Their Projects.*]

After our briefing, we all donned hard hats, and Dave drove us to the first construction site hidden behind a high solid board fence. In the yard were various pieces of large equipment—cranes, generators; a portable toilet marked "women" (there are women engineers on the team); and a small unoccupied portable office. All the while, Dave kept up a constant spiel, telling us little stories about the problems, progress, and people working on this enormous job. [*Lesson 3: Effective Project Managers Are Good Communicators and Are People-Oriented.*]

The immense hole, like an open mine shaft, was 50 feet in diameter and 140 feet deep. The walls of the shaft were gray and had a stippled appearance where the shotcrete had been applied. Except for street noises nearby, not a sound came from the site. On one side of this huge crater was a narrow porch and a mesh construction elevator about three by five feet and made completely of mesh grillwork with trap doors in both the ceiling and the floor. Dave said that the elevator had faltered recently when the director of transportation (clad in a good suit) had been in it. The director had to go out the upper trap door and climb up a ladder along the sheer wall of the shaft. This afternoon, however, the elevator was on its best behavior and chugged obediently to the bottom of the pit. [*Lesson 4: Effective Project Managers Aren't Afraid to Get Their Hands Dirty and to Visit the Troops in the Field!*]

As we walked through the dimly lit area, Dave exchanged small talk or gave encouragement or instructions to each as we passed by. [*Lesson 5: Effective Project Managers Have Good Personal Relationships with Their Workers.*]

This section of the tunnel, empty but for a few pieces of equipment, was almost finished. The area was neat and clean with no debris such as that found topside, and, except for a little condensation on the ceiling, there was no evidence of water—no leaks in the smooth finished walls of the tunnel. From the shaft we walked through the tunnel along the trackless brick and mortar roadway. As we rounded the curve to the station, we half expected to be met by one of those speeding, sleek modern cars that make up a Metro train, and we anticipated the thunderous rumble as it pulled up parallel to the 600-foot station platform, which is exactly the same length as a six-car train. Just below, exhaust vents are located strategically to cool motor and wheel units. Dave explained some of the subtleties of the design, such as the cantilevered passenger loading platform. Should a person fall from the platform, there is room underneath to roll to safety. Another interesting feature is placement of a vertical shaft where air pushed along by a train is exhausted upward to prevent a sudden pressure increase in the station. [*Lesson 6: Effective Project Managers Are Ever Mindful of the End User.*]

Dave showed us the inner workings of the partially completed Metro station—the room where all the electrical and control equipment would be located, and the holes where escalators and elevators would be installed. Then we retraced our steps and returned to the surface. [*Lesson 7: Effective Project Managers Are Willing to Describe Their Projects Even in States of Incompletion.*]

Our next stop was to see an "American Method" shaft. The 200-foot-deep shaft looked like the inside of a barrel, with hoops and barrel staves. After a short elevator ride, we stepped into an area in sharp contrast to the previous tunnel—wet, muddy, damp, cold, and noisy. Large earth moving machinery lumbered about as men dressed in hard hats and heavy-toed boots worked at various tasks. Again, it was evident that Dave was no stranger here. He joked good-naturedly with the men and answered a few questions about the work at hand before moving on to a NATM area. [*Lesson 8: Effective Project Managers Are Always Willing to Give On-the-Spot Instructions!*]

As we moved into the tunnel, we saw the various stages of the "Austrian method" section of construction—the shotcreted surfaces; then the fleece application where the material similar to a carpet pad covered the walls of the tunnel. The fleece acts as a "wick" to extract water and drain it to the sublevel of the tunnel below the tracks. The next stage, the vinyl covering put over the fleece to provide a dry surface for the final layer, is similar to present-day swimming pool construction. When we returned to the elevator, it was busy—a workman was bringing dinner down to the workers below; they would continue working through the shift. Dave showed us the surface work going on near a key station in downtown Rockville, Maryland. He related several stories about his relations with the local property owners—problems that were smoothly resolved with tact and diplomacy. He also told us about alternate designs that had been worked out in real time to resolve interfaces with present and future buildings and to cut costs. [*Lesson 9: An Effective Project Manager Delegates Authority, Lesson 10: Considers the Public Good,* and *Lesson 11: Works Out Problems in Real Time.*]

As discussed in Chapter 2, the Metrorail system in Washington is widely considered by architects, engineers, and (most importantly) its riders to be an excellent major transportation system that provides a much needed service to the public in a thoroughly modern, efficient, and well-run environment. Although the Metro project did experience cost growth from its original estimates, principally because of specification changes, it was adequately funded throughout its construction phases. But plenty of money alone does not make a successful project. Its management credits on-site project managers like Dave Shilling as a principal reason for its construction successes. We noticed that experienced project managers were chosen with great care and given full responsibility *and* authority for cost, schedule, and performance of their jobs with little interference from outside. The project manager was the chief engineer, chief administrator, and chief public relations officer for his portion of the work.

One very impressive technique that Dave Shilling used to help manage the project was use of a Board of Engineering Consultants. Technical experts from industry, government, and foreign countries who make up this board met once a month for two to three days to review design and production progress in detail. This "peer" review provided the project manager with a detailed, searching, critical, objective review of problems and progress, and recommendations and suggestions for improvements, cost savings, and problem resolution. Dave Shilling made good use of this board because he could quickly change the direction of the job to take advantage of potential cost savings without having to go through a bureaucracy or complex approval cycle. He also had a value engineering clause in his contracts with construction firms which split the savings 50/50 between the contractor and the government. This arrangement proved to be a good incentive and has resulted in numerous cost savings as the work progressed. In essence, Dave teamed with his contractors to save money for the taxpayer which, to us, is a win/win situation.

So the project manager's team consists not only of project office and project field personnel but of advisory boards, consultants, and the contractors who do the work. Larry Seggel, the Army MLRS project manager who was quoted in Chapter 2, also relies heavily on his contractor team to provide excellence in the project. Not only did Larry's personality come through when we talked to him, but it was easy to see that mutual respect and trust in a competitive environment, rather than an adversary relationship, pays off in weapons projects. We continue the interview with Larry. He was speaking of LTV, the prime contractor, and had just been asked, "How can we relate your experience with other programs?"

SEGGEL: *The contractor (LTV) had that same business pride in doing it that we did. That crazy bunch of Texans, you just couldn't tell them they couldn't do something!*

PALADINO: *I think it goes back to the fact that the requirements were stable, and everybody had a job to do, knew what it was, and they were allowed to do a good job!*

SEGGEL: *We had funding stability: we had requirement stability; and we had acquisition strategy stability. Everything stayed constant from the time we started.*

Commentary. We have made a point about the need for stability early in the book, and we will remake this point several times. One cannot help but believe that Larry Seggel himself had a major influence in achieving the program stability. Time and again, Larry would have to discourage the innovators from over-improving and meddling with the MLRS concept, which itself was based on simplicity. We can conclude that strong project management, exercised throughout the project's lifetime, is one of the essential elements required to achieve program stability. To continue with the interview:

RDS: *. . .The initial selection of the two competing contractors. Were they selected from other contractors?*

SEGGEL: *Yes, they were selected competitively.*

RDS: *How many, say ten, maybe?*

SEGGEL: *No, there were three.*

PALADINO: *There were six in the original concept definition study, but then there were three that bid.*

SEGGEL: *There were three that bid the development.*

RDS: *So it was a narrowing down process to get the good contractors. You know this is one problem now, apparently, with the competition rules and laws. Some people are telling us that you're opening up the competition, making it wider and wider as you go downstream instead of narrower and narrower; and therefore you're getting a chance to add some bad eggs. But it was narrowed down from six to three and then to two?*

SEGGEL: *Yes, that's right. And then to one. And then we were prepared to open it back up again if we had to in production but didn't need to because they* fulfilled all their obligations *and then some.*

RDS: *This was a way of keeping some competition?*

SEGGEL: *Yes! The leverage was there all the while. Fear of competition even if there wasn't competition.*

PALADINO: *We used a strategy of validation of two competitors. Then we had a maturation phase with one.*

SEGGEL: *In today's terms: [maturation means] full-scale development.*

PALADINO: *We had competition between those two guys . . . they first finalized the design and, through a later competition, we started production for the first four years. And then when we got into full-scale production, we had the choice of going to a second source or a multiyear procurement. We released an RFP to this one single contractor both ways. Give us the multiyear bid and give us your price for the educational portion [education of a second source] if we develop a second source. So he [the contractor] had the competitive pressure even though there was no competitive proposal or contractor involved, he had the pressure with Congressional language mandating, eventually, a second source if the multiyear procurement didn't pan out. So we had competition in the three separate phases of the whole acquisition strategy.*

Commentary. Despite the intensely competitive environment that existed at first, and that was stimulated throughout the project by retention of a second source, it was apparent to us that the government and the contractor considered themselves a closely knit team. Both the contractor and the government project office personnel were excited about the job they had to do, and a lot, maybe even most, of this high motivation level came directly from the actions and the personality of the project manager. Therefore, it was not so much the procurement strategy that created success but the maintenance of a team spirit where each team member identified with the overall project goals and was willing to go to bat for them and stick with them. To continue:

RDS: *What type of financial reports did you get from the contractor? Did you get standard reports?*

SEGGEL: *Yes! Standard cost performance reports.*

RDS: *Did you find that that helped you determine where you were, or was this just a sort of paperwork requirement that was not used?*

SEGGEL: *Of course, when we say standard we're talking about the Army CSCSC [Cost/Schedule/Control/Systems/Criteria] reports, top performance reports. It tells you where you are and where you're going and how you track it.*

RDS: *Do you attribute that to part of your success? Or do you think that's just more paperwork?*

PALADINO: *The overall financial management method resulted in part of the success. We believed in making do with the dollars we had, and if there was a change that required that we spend some dollars we would go find some place to remove some dollars from the program* and make the program succeed with the dollars we had [emphasis added].

RDS: *Is that a key point?*

PALADINO: *Yes. We didn't always go back and ask for more money every time something came up. What can we do without or what can we do different and still meet the budget? If the contractor came in with some kind of overrun we would say "Tell us where we can take the scope out of the contract without hurting the program. Give up a 'nice to have' or something like that." We constantly had a forward look at where we stood on things. . .*

SEGGEL: *And we were able to act such that the crash didn't occur. We were ahead of the game. General Moore [then Commander of the Army Missile Command] would say [that we were] "proactive."*

Commentary. The foregoing illustrates that a project manager can manage within funds provided he or she is willing to make trade-offs within the program. Again, Larry's strong personality, his backing by management, and his long-term commitment to the program were significant factors in achieving stability. We sensed that Larry had control of his program and was not bandied about by superiors, contractors, and government bureaucrats—or at least he did not allow it. Commitment to live within a dollar amount really brings to the surface what management is all about—the ability to make trade-offs resulting in win-win situations.

RDS: *Did you get ideas from the Europeans too?*

SEGGEL: *No, not so much in that program or development but it was because their involvement was essentially financial. They gave us some money too and then it was information transfer. But the fact that they were participating, and that they had money involved and that they were signed up with us for this program, I've got to suggest, had some influence on the stability and the constancy of the U.S. funding and the U.S. support. We had, then, national obligations we had to meet.*

RDS: *So the international obligation was beneficial?*

SEGGEL: *Yes, that supplied some leverage as well. It suggests that while I might be inclined to cut the program $5 million, it's going to have an impact on the [international] program. I'd rather not do that and have Germany and France and the U.K. get all upset. So even my superiors had obligations to somebody [the Europeans]. And, as a result, all of those things, I think, played together to the benefit of the program.*

Notice that the project team carefully selected the best contractor through a weeding-out process, narrowing down from six study contractors to one. Over the years, the project team developed a very close relationship and a very good rapport with this single contractor. The project team had the option open to go to a backup contractor if relationships were to sour with the prime. (An ex-project manager with the same agency told us later: "This would be extremely difficult to do once you go to one contractor. First, the other guy may say, 'No.' Second, it would cost a ton. Third, the schedule would be prohibitive.") But for now, suffice it to say that the project team developed and carried out a very systematic and organized "acquisition strategy" designed to provide the best industrial team participant. Second, good financial management reporting and cost control throughout the project, as well as a management commitment to keep out unnecessary changes, permitted the project office to stay within its resource allocations. Third, the project had foreign government involvement, which was good not because of technical expertise provided, as in the case of the Metro project, but because of the financial commitment of and to foreign governments which enhanced project stability. All of this was done in an atmosphere of camaraderie and team enthusiasm—sparked by the project manager.

We were curious about the project team for MLRS. What were their backgrounds and motivations? Why was it considered by the Army to be one of their best project offices? On this subject, the interview with Larry Seggel concludes as follows:

SEGGEL: *Our project team here has stayed together.*

RDS: *How was this done?*

SEGGEL: *Leadership. Guys like Rich and I have been here since the start. I've been with it since it was an idea—1975—before it was a project. Rich came on in '76. The chief of engineering came on in '76 so he was on the "borning" team early.*

RDS: *Rich, what is your title?*

PALADINO: *Chief of Project Management . . . on the financial end.*

SEGGEL: *Rich heads the Project Management Division of the project; his business is financial—acquisition strategies and schedule. Our chief engineer has been here since day one.*

Commentary. Note that members of the project team other than the deputy project manager also had continuous service. This team appears to have resisted the tendency to want to rotate personnel and to change the team's composition periodically—without regard to the impact on the project itself. Yet, clearly, some changes and additions in skill categories and skill levels could be beneficial as a project proceeds through its various phases.

RDS: *What is your background, Larry? Engineering?*

SEGGEL: *Where'd I come from? Matter of fact, we're both* industrial engineers by degree. [emphasis added] *I guess that tells you something about why we feel the way we do about business and money and things like that because we're trained as engineers with those disciplines. Also, in our case, while you have an engineering degree you get a whole lot of management and you even get some law. So that combination of things is, I think, highly suited for the kinds of project management jobs as contrasted to, say, mechanical engineers or aerospace engineers who don't get the business aspect of things other than by taking other courses or through osmosis. I came up through the project business. Before I came here, I was the chief engineer on the Lance.*

Commentary. Technical as well as management and business expertise existed at the top level throughout the MLRS project. These skills were embodied in the top two civilian managers. Both had an appreciation for the engineering requirements of the project but were also integrally aware of the business and financial impacts of their technical decisions.

RDS: *How many people do you have on your team?*

SEGGEL: *Well, when you talk about civilian and military both, and our European contingent here working with us (we have European guys right on our team working as integral staff members), we've got about 120 people. And it takes every one of them.*

RDS: *These Europeans are engineers?*

SEGGEL: *They're professionals in their area. We've got configuration managers; we've got test people; we've got military; and we've got logisticians. They're here as part of the team and their function is to bridge between their country and ours in their area of interest. But they work just like us. We don't treat them any different.*

RDS: *Do they report back to their own governments?*

SEGGEL: *No, they report to us, and they work for our guys just like a U.S. civilian would or a U.S. military would. We direct them in all regards, but their duty to their country is to be a pipeline of information. They can talk to the folks back home in their language about what the test program is doing and how that's going. But that's an incidental function to what they do here as testers. We give them the same assignments as our people. Fact is, we just don't care where they're from and they have done great! I've got nothing but good things to say about my partners.*

RDS: *Your staff is probably over 50 percent engineering?*

SEGGEL: *In our organization? Yes, considerably more than that; I think probably 70 percent.*

RDS: *There are some services and other agencies that have projects where the project manager is either an administrator or businessman. He may have both technical and business people on his staff, but he depends principally upon the contractor for the detailed technical expertise.*

SEGGEL: *Not so here! Not so here!*

RDS: *Do you need to have the detailed engineering background?*

SEGGEL: *Yes! You ought to be able to understand the turf; otherwise you can get your backside sandpapered real good. And we manage in detail. In all areas we're very tight with our contractor. We know what he's doing, when he's doing it, and why he's doing it!*

RDS: *Do you have anybody located with the contractor?*

SEGGEL: *Yes. This is going to surprise you—we have three people that reside with the contractors. Other projects have big field offices. I don't like them. Big field offices have a tendency to wind up developing a head and an existence of their own. And because he's there and convenient, the contractor talks to him. Because the contractor comes in wringing his hands, 'I've gotta have a decision now,' he makes a decision, and back here we may or may not ever know what went on.*

RDS: *Do they begin to side with the contractor?*

SEGGEL: *More importantly, they begin to act independently from home base. So I just don't have that. I don't give them enough wherewithal to do that because I'd rather my team right here be in control. So I don't believe in large field offices.*

RDS: *These three people you have, are they technically oriented?*

SEGGEL: *One is an engineer and the other two are military. They also have a secretary. . . .*

RDS: *So you have a mixture of technical and business people on-site?*

SEGGEL: *Yes, but you know they haven't enough time or enough expertise to do anything that would hurt us out there. They have to function as eyes and ears.*

RDS: *They can do that effectively?*

SEGGEL: *Absolutely. They do a good job of that and that's exactly the function I want. Now if I gave them three more there'd be more than eyes and ears and they'd start doing things. . . .*

RDS: *Start being a bureaucracy?*

SEGGEL: *Yes! We acted the same way on Lance and both programs wound up with good weapons in the field. I tend to feel like I'm more right than wrong.*

RDS: *Dr. Rees told us, on the Apollo program, that he had engineers out in the contractor's plant.*

SEGGEL: *Well, we do it the other way. Home's here and keep your toothbrush packed because you never know when you're going to be jumping a plane to go. That's one of the things that makes me so proud of this bunch of people that work here, I don't have to tell them to go somewhere. They say, "If the son of a gun doesn't look just exactly the way it ought to be—I'm going." It's that kind of attitude about things that's so super, I don't have to whip them to get them to go.*

The predominant message that comes from the foregoing is that there was good stability, and there was both business and technical expertise in the project team. Many team members worked on the very early laboratory development of the weapon system and are still with the team. Keep in mind that Larry's title is *Deputy Project Manager*, not Project Manager. The project management position had been held by a number of Army officers during the program—about seven project managers during the first ten years of the program. This means that the military project managers spent an average of less than a year and a half each on the program. As a result, the continuity of project management resulted from a strong civilian deputy project manager who stayed with the project, and a mostly civilian project management staff. Larry and his right-hand man, Rich, are both industrial engineers by training, and their staff's background is about 70 percent engineering. The project keeps in close touch with the prime contractor through a very small field staff plus frequent visits by home project staff members. (In construction projects, the project manager should be at or near the site of the major effort.) Larry is not in favor of big on-site staffs because it insulates him from the real world. The on-site people function as his eyes and ears in the contractor plant. Larry is careful to point out that what worked for MLRS won't necessarily work for other projects, but we can tell you from our own experience and from reviewing other successful projects that continuity, technical competence, an "in-touch" attitude with the customer, and strong leadership don't hurt.

The Qualities of a Project Manager

It is a common view of those outside of government that people in government are somehow different from the rest of us! An all-too-common view is that people in government (1) have little regard for the taxpayer's dollar; (2) are sluggish, unresponsive, and inefficient; and (3) are incompetent, inadequately trained, and overpaid. This unfortunate view of the public servant has been foisted upon us by cartoonists, humorists, columnists, and (occasionally) whistle blowers, all of whom are trying to make some special point without realizing the overall negative effect of their sketches, actions, or words.

In truth, the people in government are very much like the rest of us. They want to do the very best job possible for the taxpayer, they are honest and hardworking, and they see their role as an important service to our country. Indications that there are many competent people in government lie partially in the fact that even those without special inroads to special projects, after leaving the government, are usually very highly sought after by private industries, universities, not-for-profit corporations, and research institutes. Proof also lies in the fact that many previous industry leaders, scientists, and administrators were hired by the government from the private sector to transfuse business expertise from industry into government. The people who move from government to industry or vice versa are the same people before and after the move. That is, their basic values are the same, but their outlook and perspective on key issues can be vastly different. As a result, one notices a change in behavior in people who have made the move into the new environment. Why does this change occur? Strong influences in this change in behavior are the constraints, rules, regulations, and policies encountered when moving from the private sector to the public sector. In order to achieve megasavings in megasystems, the entrepreneurial spirit, drive, and ability must be preserved after the move from industry to government. The same management philosophies that make excellent corporations work can make excellent projects work.[2] And the most important person who must acquire and retain these attributes is the project manager.

Whether he or she comes from industry or has come up through the government hierarchy, the project manager must have certain qualities that make the whole thing click. The Department of Defense Authorization Act of 1986 placed new requirements on project managers' qualifications, but little is said in this law about the less tangible aspects of the job, and continuity has not yet been achieved despite a heroic attempt by Congress to try to stabilize and lengthen the tours of duty for project managers.

What is the key quality in a project manager that sets the stage for success in a large project? Assuming that the project manager has the required dedication, persistence, patience, vision, drive, determination, and enthusiasm, and the qualifications for the job (education, continuity, and experience), what are the critical factors that distinguish the one who can successfully direct a megaproject?

In a presentation to the China Association for Science and Technology in 1986,[3] Dr. Lewis R. Ireland pinpointed four critical success factors for large projects: (1) *proactive leadership*; (2) owner involvement in the project; (3) initial planning and follow-on planning; and (4) life-cycle management. The critical success factor that has the greatest bearing on the attributes of the project manager is *proactive leadership*! To understand what proactive leadership is, we first have to know the definition of *proactive*. The 1981 edition of *Webster's New Collegiate Dictionary*, defines "proactive" as "involving modification by a factor which precedes that which is modified." In a project, it means foreseeing a problem and avoiding it through preventative measures rather than waiting until the problem occurs and then trying to fix it. In simpler terms, a proactive person is one who doesn't wait for the roof to fall in but anticipates it and does something about it. It is management by objectives rather than by exception.

In each of the excellent projects we investigated, the project manager exercised proactive leadership. Although we have known some project managers of good reputation who have spent most of their days "putting out fires," the really excellent ones are those who have thought out their moves first, have a plan of action, and stick to it. The excellent proactive project manager is much like the master chess player who thoroughly thinks out his moves, *and the alternative possible moves of his opponent*, before making a move.

Proactive Leadership

Adams, Bilbro, and Stockert, in a publication prepared for the Project Management Institute,[4] list five key components of what they term "developmental leadership":

1. A willingness to manage,
2. A positive orientation to people,
3. The skill to communicate mission and purpose,
4. A decision-making technique that builds commitment, and
5. A leadership profile that creates support.

To make these five points fit into a definition of proactive leadership, we merely need to add a sixth component: vision, which is an ability to picture the end result as it will be when completed.

Adams et al. describe a project manager's personal attributes that correspond to each of the first five components. For example, under "Communicates Mission and Purpose" they ask, "Is the project manager persuasive?" "Does he communicate mission and organizational values?" and "Does he listen without bias?" They list a total of sixteen questions that should be asked to assure that a project manager has the first five qualities. These questions, joined with a series of questions relative to the *visionary* makeup of the project manager's character, would form the basis for an excellent questionnaire to determine a person's suitability as a proactive leader. The questions on the visionary qualities of a project manager might be couched as follows:

1. Can the project manager envision the end result as it will have been accomplished?
2. Does the project manager possess the ability to distinguish rapidly the important points in a problem, plan, or discussion?
3. Is the project manager open to innovation and imaginative solutions?
4. Does the project manager employ synergy and synthesis in creative problem solving?
5. Does the project manager have the ability to see how a solution can arise out of the depths of a problem?
6. Is the project manager capable of maintaining his or her vision, dedication, and commitment even while overcoming problems or failures?
7. Does the project manager have the ability to bring people together in a mutually constructive environment to produce a desirable end result?

Attributes of the Successful Project Team

Commitment of the entire project team, *including* the project manager, to critical success factors (owner/user involvement, planning, and life-cycle management) are also necessary for excellence in managing megaprojects or acquiring megasystems. A critical success factor is defined more fully by Cleland and Kerzner[2] as "those characteristics, states, or variables that exist or can be properly sustained and managed which will have a significant impact on the attainment of the project objectives on time and within budget." The first critical success factor, owner or user involvement in the project, was visibly evident in the excellent sample megaprojects we studied in space, civil, and military acquisition (the Apollo program, the Washington Metrorail system, and the Multiple-Launch Rocket System [MLRS]).

In the Apollo program, the "users" were considered to be the astronauts. Astronaut involvement was intense during the early definition phases; throughout the detailed design and development phases; and in the production, operation, and training phases. Astronauts attended design reviews and detailed technical meetings and were represented on interface control boards and flight operational readiness teams. They were provided with the complete status of critical program elements and given free access to analysis results, test results, interface studies, and program reviews on all of the Apollo systems and components. Astronaut sign-off was required for critical manned spaceflight related activities. Key astronauts kept themselves on an almost constant tour of the United States in their own jet aircraft, visiting the various NASA centers, prime contractors, subcontractors, and launch sites. They expressed concern when safety problems arose and were listened to when they gave advice in design reviews, pre-flight reviews, and post-flight debriefings.

The end users in the Metrorail project are the thousands of daily riders. In user-oriented projects like the Metro, user advocates or surrogate user groups are set up early in the project to provide inputs relative to comfort, safety, user-friendliness of ticket machines, and so on. In the case of the Metro, as described earlier, the very best experience of other rapid transit systems was tapped through visits, interviews, and design reviews. Special consultants and advisory groups were enlisted to ensure that the end user would have available a neat, clean, attractive, efficient, and safe system of transportation.

The Army's TRADOC was the user advocate in the case of the Multiple Launch Rocket System development. (For larger systems, the Army's Operational Test and Evaluation Agency is the major voice representing the user.) This command represented the interest of the GI who would have to operate and maintain the system. User advocacy is particularly important in military systems because development engineers tend to become enthralled with high-technology systems that may not be practical for operation or maintenance by the typical trained enlisted soldier under battlefield conditions. Project teams that ignore the safety, comfort, and ease of operation and repair sometimes find themselves the target of whistle blower groups, weapons system scandals, and after-the-fact Congressional investigations. (See Dina Rasor's description of her ride in an M-1 tank in the book, *The Pentagon Underground*.[5])

Critical success factor number two is detailed planning. Just as mega-planning is required by the federal government to integrate and time-phase the acquisition of all major systems, micro-planning is required by the project team to assure that all pieces of the puzzle—hardware, software, documentation, training, spare parts, logistics, and maintenance—fit together. One-time planning is not enough. Plans must continually be updated, revised, and amended to take into account day-to-day, week-to-week, and month-to-month changes caused by problem areas, improvements, cost savings, interface anomalies, test results, feedback, and operational evaluations. The project team is constantly engaged in a planning, management, and replanning activity to assure that the timing of project elements coincides with the needs. High-tech computerized networks and critical path bar chart computer programs are available that will make replanning and re-estimating easier than manual methods. Once baseline schedules and dollar budgets are put into the computers, they can easily be updated to take care of small perturbations in availability or requirement dates and times. The project team must have the skills and the tools as well as the commitment to continue this planning and replanning effort throughout the project.

The third major critical success factor for management of large projects is *life-cycle management*. Procurement of a system or project is only the first step in ensuring its availability and effective use during its lifetime. Life-cycle management means that the initial design must take into account the future needs of repair, maintenance, training, operation, logistics, and eventual upgrading and even disposal of the system. Continuity and commitment to owner or user needs throughout the system's lifetime means continuity in personnel, support, and funding. The project management team itself must have a long-term charter and an established and capable staff that can stick with the project until its completion. Because of the increasing time-span of major projects and systems—ten to fifteen years—any single team member may end up having two, or at the most three, project team assignments within his or her professional career if true team continuity is achieved. This leads to the increasingly evident fact that project management is a profession in itself. No longer can a project team for a major system development be put together on short notice without adequate consideration of the tenure, future career path, skill mix, or long-term availability of its personnel.

Just as a commitment from the President and Congress is needed to provide consistent and constant funding, personnel, and moral support to the project throughout its life-time through the establishment of national goals and a dynamic national plan, a long-term commitment is required by the project manager and the project management team. In Cleland and Kerzner's paper on "The Best Managed Projects,"[2] the word *commitment* appears profusely. Successful projects are those in which the "owners" exhibit a "critical *commitment*" that the job be managed properly. Success results from "*everyone's commitment* to the project completion dates." In periodic project coordinating meetings, project personnel in attendance must be "*committed* to facing and resolving problems with prompt decisions." Companies "*committed* top management to these meetings" (in addition to project personnel). "The importance of *everyone's commitment* to project completion dates cannot be overemphasized," and "*commitment* to quality at all levels and elements of the organization" were distinct characteristics of successful projects. And, finally, the "development of a *project commitment* and sense of teamwork" was evident in excellent projects. In their own in-house study of excellent projects in fifteen corporations, Cleland and Kerzner found that "*executive commitment*" was a key driver in obtaining outstanding project success.

Within the framework of total commitment to excellence in achieving scheduled dates and providing a quality product, many other features of successful project teams were cited: comprehensive strategic planning, realistic and thorough project definition, aggressive (proactive) leadership, regular surveillance, updating of project skills, careful selection of key personnel, frequent use of independent task teams, and computerized work scheduling systems.

Systems Engineering

A principal tool used by the project manager through either in-house or contracted work to remove uncertainty is the discipline of systems engineering. Systems engineering, as defined by Blanchard and Fabrycky[6] is "a process that has recently been recognized to be essential in the orderly evolution of man-made systems. It involves the application of efforts necessary to: (1) transform an operational need into a description of system performance parameters and a preferred system configuration through the use of an iterative process of functional analysis, synthesis, optimization, definition, design, test, and evaluation; (2) integrate related technical parameters and assure compatibility of all physical, functional, and program interfaces in a manner that optimizes the total system definition and design; and (3) integrate performance, producibility, reliability, maintainability, supportability, and other specialties into the total engineering effort."

The following are key systems engineering disciplines that, if systematically and rigorously applied by project managers and their teams, can and will result in the type of exact definition needed to procure systems, in phases, at a predictable cost.

1. *Requirements Analysis.* This discipline is the very first and the most important step in major systems acquisition. It is performed through an iterative, continuous interface between the users and the potential suppliers of a system. It employs the discipline of systems analysis, which seeks to develop an inseparable bond between what is required and what can be done. The systems analyst must have an in-depth knowledge of the technology of the system to be acquired as well as an ability to communicate realistic potential capabilities and performance parameters to the potential user or user advocate. Then, the systems requirements analyst must be able to interpret user requirements and, through an iterative process, to develop the specifications of a system that meet these requirements. No other activities in the systems acquisition process can take place until preliminary requirements are defined. Lack of specificity in system requirements, and changing of these requirements during programs, has been one of the major reasons for cost growth, overruns, and schedule slippages. The systems or program analyst, who must be knowledgeable and capable in both technical and business skills, is a key player in this first, most important, and vital step in systems engineering.

2. *Engineering Analysis.* Once preliminary requirements are defined, an engineering analysis of these requirements results in preliminary detailed system specifications. This includes a study of the effects of the user and the operational environment on the system and of the interactions between major subsystems. This discipline can feed back to the previous step to adjust preliminary requirements if necessary. Disciplines included are human factors; thermal, shock, vibration, pressure, and other hostile environments; interfaces with other systems; affordability; maintainability; and logistics.

3. *Computer-Aided Engineering.* As requirements and preliminary specifications become well defined, designs start emerging in the form of manual or computer-aided designs. Computer-aided design provides the systems engineer with a tool for rapid drafting, analysis, changing, and estimating of preliminary and final designs. Computer-aided design systems can incorporate data bases of material costs, manufacturing time and labor costs, and assembly times and costs to permit project engineers to evaluate the cost impacts as well as the performance aspects of a design. Once a design is generated, it can be manually or automatically translated into actual prototype or production hardware through computer-aided manufacturing techniques.

4. *Engineering Simulations.* As engineering design evolves into the assemblies, subassemblies, major components, and parts that make up a major system, computer simulations can imitate the reaction of these entit-

ies to various external forces and environments. Three-dimensional stress, thermal, and dynamic analyses that take into account internal and external forces and environments tell the engineer exactly how a part, component, or system will react under any condition. Computer pictures show the engineer how the design deforms or fails under a wide variety of conditions. Knowing how the design fails and how large the pressures, temperatures, and forces must be to induce failure help the engineer redesign the item for the lightest possible weight, the best performance, or the lowest cost. Simulations permit the "tryout" of many different materials, finishes, tolerances, circuit arrangements, software programs, response rates, and configurations in a short time without actually building the item and testing it in the laboratory. Human factors can be thoroughly considered and evaluated through simulations of machine interactions with human response rates. A design that has been thoroughly subjected to exhaustive engineering simulation will have a maximum potential of working under real conditions and will be significantly less prone to failure, redesign, or cost escalation once it is built.

5. *Formal Design Reviews.* Indispensable tools for project managers in major systems are formal design reviews. These reviews are set up at key points in the acquisition cycle, such as at the end of concept definition or at the conclusion of preliminary design. In these reviews, the project manager appoints a multidisciplinary board of experts from both within and outside his or her team to perform a detailed review of the project's status. Design reviews may also include scrutiny of hardware and software test results, production designs, documentation, and changes. In these reviews, design review boards are given complete details of the system or project's details through presentations, reports, drawings, and specifications and are permitted to submit verbal or written comments and suggestions. The design review board must then address, discuss, and decide on what action is required on each and every input.

Detailed, formal design reviews are an excellent means of surfacing potential technical or programmatic problems before they can have a significant impact on cost, schedule, or performance.

6. *Test and Evaluation.* This discipline is an integral part of any system development program. Tests are used to verify if a concept is feasible, to determine the effect of extreme environments on the system, and to determine if it has proceeded to a point where uncertainties have been removed and production and/or operation can begin. Failure to conduct tests and to thoroughly evaluate performance before starting production has been one of the major causes of performance shortfalls, cost increases, and schedule slippages. The General Accounting Office has reported that the production of some major weapons systems began with only limited operational test and evaluation results.[7] The GAO report said, "When major weapon programs do not undergo sufficient operational test and evaluation before

production, the risk of costly redesign and modification after deployment increases."

The GAO went on to give the following specific examples:

For example, at the time production began on the F/A-18 aircraft, it had not undergone critical aspects of planned OT&E [Operated Test and Evaluation] that was called for in the program schedule because of the system's immature development at that time. GAO points out that expensive retrofits were required on F/A-18 production models to correct problems identified during operational testing performed after the production decision was made. The problem of insufficient OT&E before production start-up prevailed with all the other major weapon systems included in GAO's review—High Speed Antiradiation Missile, Air Launched Cruise Missile, B-1B Bomber, and the Sergeant York Air Defense Gun.

In another recent report,[8] the GAO found that successful programs require a meticulous attention to planning, carrying out, evaluating, and reporting tests; and to the modification of systems based on these test results prior to proceeding with production or operation. Detailed attention must be given to selection of test items, expansion of test programs to include both operation and maintenance, test site selection, test personnel and training, test facilities and resources, test and support equipment, test supply support, test performance, and test reporting. One of the aspects of a test program that is vital to the project manager is feedback of test data and information. This feedback must be both rapid and accurate if the project manager is to step up to the resulting design, specification, and programmatic changes.

7. *Reliability Engineering*. This discipline is available to the project manager to assist in the projection of estimates of system reliability and to identify and prevent problems that may result in the event of the failure of a system, subsystem, or component. Failure mode and failure effects analyses are performed on all hardware and software to determine whether subsystems, parts, or computer programs should be redundant, to determine which failures might be critical, and to find means of increasing system reliability. Reliability engineers have an input to life-testing programs, statistical sampling techniques, and design reviews. They develop information on projected average times between failures and they forecast when failures might occur. These inputs are used to help the project manager determine repair, periodic maintenance, and spares support requirements.

8. *Quality Engineering*. This important discipline is a close cousin to reliability engineering in that it evaluates all program technical and performance factors to assure the project manager that a high quality system will emerge from development and production. Quality engineers plan pro-

duction sampling and inspection techniques; specify tolerances, materials, parts, and methods where these variables may affect system quality; and participate in engineering, design, testing, and production reviews.

9. *Human Factor Engineering*. Human factor engineering includes ergonomics, maintainability, safety, user visibility, acoustics hygiene, and user comfort. Human factor engineers develop operational sequence diagrams and detailed task analyses to establish and verify manual operations; automatic operations; operator decision points; operator control actuations or movements; transmitted information; or received information using indicator displays, digital readouts, and video or audio stimuli. The comfort, safety, and efficiency of the pilot, driver, passenger, or maintenance worker are foremost in the priorities of the human factors specialist. These specialists have played an important role in such diverse projects as EPCOT Center, the Washington Metrorail system, and manned spacecraft.

10. *Configuration Engineering and Interface Control*. The complexity and number of subsystems in modern projects and systems has spawned a whole new engineering discipline to assure the project manager that all parts of the system will fit together properly when the whole is designed and put into operation. This discipline is called, variously, configuration engineering, configuration management, interface control, or change control.

When one part of a complex system is changed, it usually has an effect on one or more of the other parts. For example, changing the shape or weight of a component could affect the location of the system's center of gravity, its response under vibration and shock, or its balance. Changing a computer circuit can affect the programming of software, and changing flight hardware can cause changes in the design of testing or handling equipment. The configuration engineer maintains a complex change tracking system and an approval record to be certain that all impacts of a change have been reflected in the design or operation of interfacing hardware, software, or documentation. Systematic, rigorous tracking of changes and change impacts is vital in all major systems, and the configuration engineer sees to it that this tracking is accomplished.

11. *Design-to-Cost Engineering and Life-Cycle Cost Engineering*. Cost can be the fixed quantity in the cost/schedule/performance equation. The design, development, and production of major systems to stay within unit cost targets has resulted in the need for design-to-cost engineering; the acquisition of systems to life-cycle cost targets has spawned life-cycle cost engineering.

Rigorous application of systems engineering disciplines such as those described here is a must if successful projects are to result. The project manager must exercise leadership to assure that these and other required systems engineering disciplines are given proper attention throughout the project.

For DOD: Civilian or Military Project Managers?

There are two schools of thought concerning project manager selection for Department of Defense weapons systems and other major DOD projects. One school of thought, which has been the predominant one to date, is that the project managers for military projects should be military officers. The argument is that a military officer can be a user advocate for his or her own project because he or she has been trained in tactics, strategy, and the use of weapons. Despite a DOD Directive (*5000.23: System Acquisition Management Careers*, November 26, 1974) that states, "Personnel should be selected on the basis of skills and experience. . . regardless of military or civilian status," the general policy of the services is to choose military project managers. In fact, the memorandum that accompanied the DOD directive when it was issued stated that the directive's thrust is to develop a cadre of *military project managers* and that assignment of a civilian as a major project manager would occur only *"in a case of extreme circumstance"* [emphasis added].[9]

The second school of thought—that military officers should be trained and used primarily, perhaps even solely, as field commanders who know and have experience in military strategies and tactics rather than project management or administration—is expressed in detail in Gary Hart's book, *America Can Win*.[10] Hart's view is that there are too many officers and that they are being trained, through formal education as well as through on-the-job training, as office managers rather than field commanders. Hart says that the educational establishments for military officers "must focus on war, not management."

The answer to the choice of a manager with a green suit, white suit, blue suit, or grey flannel suit (Army, Navy, Air Force, or civilian) falls somewhere between the two extremes. In reviewing successful projects, it was found that *continuity* of project management (regardless of the color of suit) was a critical factor in outstanding projects. In military projects, this continuity must come either from the military project manager, the civilian deputy, or from the project office personnel which can be any mix of civilian and military. The difficulty in providing continuity with a military project manager is that the military has several policies that usually prevent the military project manager from staying with the project throughout its lifetime. The Army Tactical Missile System[11] had four project managers in two years. The Air Force Small Intercontinental Ballistic Missile[12] had three project managers in the first twenty-seven months. Each of the first two managers was in charge just a little over one year. Policies that include the frequent movement of project managers in location and/or assignment to enhance cross-training and periodic contact with troop unit duties, the "up-or-out" promotion scheme, and the policy of assigning military personnel to jobs commensurate with their grade or rank generally prevent continuity of military project managers. Despite stated DOD policies (DOD

Directive 5000.23) that require a tenure for the military project manager for a major program of "not less than four years or until the completion of a major milestone," military project managers seldom stay in their jobs for more than two years. Promotions cause the movement to higher level jobs; retirement shortens job tenure; and assignment less than four years ahead of a major milestone—all result in projects having three, four, or even more project managers during their lifetime.

Reduction in the size of the officer corps and restriction of officer duties to combat preparation activities, although an admirable objective, is not realistically achievable in the near future. The officer corps has some excellent project management material, just as does the Civil Service work force. The answer to achieving continuity in project management, then, is not to eliminate but to firmly establish project management as a career path in the military as well as in the civilian service and to select only the best project managers, military *or* civilian, to direct large projects. Project managers must be chosen based on their outstanding capabilities and experience. Life-cycle planning and estimating must be accompanied by life-cycle project management. Rotation policies must be rescinded to permit those who have sufficient tenure to see a project or project phase through to its completion.

Life-cycle project management can be achieved through a phased approach where the project manager is assigned for one or more *full phases* of the program (program initiation, demonstration and validation, full scale development, production, deployment, or operation) *with sufficient overlap of project managers to assure full program continuity*. This approach provides continuity while permitting some progression of project management personnel. A GAO report entitled *Strengthening the Capabilities of Key Personnel (in DOD) Systems Acquisition*[9] recommends "commitment by the project manager to stay with the program through the achievement of some concrete result (such as hardware baselining solution or user test and evaluation). This arrangement would strengthen project manager accountability." Commitment *by the project manager alone* is insufficient to assure his or her tenure. A policy of life-cycle or program-phase management will have to be implemented from the top by DOD and the agencies with the approval of the President and Congress.

Other interesting observations that substantiate our findings about project managers for project excellence appear in the GAO report[9] on strengthening acquisition personnel:

> *Service management and panels believed that selection of project managers should be based on performance in the* acquisition career field. [emphasis added] *Selecting project managers from outside the acquisition career field undermines the credibility of acquisition career programs and the ability to attract promising personnel into the field.*

> *Some service managers believed that a technical educational background (engineering*

or physical science) was desirable. A technical background was seen as allowing the project manager to converse knowledgeably with functional managers and contractors. Air Force management expressed the view that the optimal educational background [for project managers] was an undergraduate degree in engineering or a physical science and a graduate degree (usually obtained after entering military service) in management.

The foregoing comments confirm that program, project, or systems management is, indeed, a career field that requires long-term, specialized training and experience, including a mix of technical, business, and management disciplines.

The Department of Defense and the services have nineteen schools located at various installations around the United States as part of their Defense Management Education and Training Program. These schools offer over 400 workshops, seminars, undergraduate, and graduate courses, many of which pertain to acquisition management. Project management and related courses for potential managers of military programs are available at the Defense Systems Management College at Fort Belvoir, Virginia; the National Defense University in Washington, D.C.; the Air Force Institute of Technology in Dayton, Ohio; the Air University, Leadership and Management Development Center at Maxwell Air Force Base, Alabama; the Office of Naval Acquisition Support's Acquisition and Logistics Management Training Center in Washington; the U.S. Army Logistics Management Center at Fort Lee, Virginia; and the U.S. Army Management Engineering Training Activity at Rock Island, Illinois. Other institutions, such as the Navy Acquisition Training Office at Norfolk, Virginia, are coming on-line on a regular basis due to the increased emphasis on and need for excellent acquisition training.

The key course for DOD and other government project managers is a 138-day course titled "Project Management" given by the Defense Systems Management College at Fort Belvoir, Virginia. This course, coupled with "substantial" program office experience and a minimum of eight years acquisition experience, is considered a must for military and civilian managers of Department of Defense programs and projects, yet not all project managers attend it. Despite the large number of military programs and educational courses, in a study of thirty-four recently appointed project managers,[9] GAO found that 100 percent of the Navy's project managers, 84.6 percent of the Army's, and 63.6 percent of the Air Force's did *not* meet these criteria. One general conclusion of GAO was that "current programs to develop military project managers fall short of those desired." GAO cited numerous changes needed in Army, Navy, and Air Force programs for training and selecting project management personnel.

Civilian project managers and deputies in DOD can attend the same courses offered to military personnel. Civilians in DOD and in other agencies and in private industry can choose a number of excellent universities which have project management and related curricula. Several universities have already established and are establishing special bachelor and masters courses in project management. These include Western Carolina University, North Dakota State University, the University of Pittsburgh, Syracuse University, the University of Southern California, the National Graduate University, Baldwin-Wallace College, the South Australian Institute of Technology, and several Canadian and British colleges and universities.

The Department of Defense, in its *Guide to Resources and Sources of Information for Acquisition Research*,"[13] lists eighty-six colleges and universities that have performed research or provided academic courses in systems acquisition or logistics. The Federal Acquisition Institute, now a part of the General Services Administration, has listed eighteen colleges that offer associate degrees, twenty-five that offer baccalaureate degrees, twenty-nine that offer graduate degrees, and eleven that offer electives in procurement, some of which include courses in project management. Western Carolina University and others offer masters' degrees in project management.

The Project Management Institute[14] sponsors workshops, seminars, and training programs in project management and related subjects. They also offer a certification program in which applicants can qualify for certification as a Project Management Professional (PMP). Each applicant must subscribe to a code of ethics; submit an education, experience, and service record; and successfully pass an examination before certification. Examination areas include human resource management, cost management, time management, communication management, scope management, and quality management. So be assured that there is no shortage of training opportunities in the methods, principles, techniques, tools, or procedures of project management; despite the fact that many project managers are not receiving the right type of training.

Project Management Fundamentals

In a 1982 document called "Principles of Project Management,"[15] NASA Administrator James Beggs wrote that a project is "a defined, time-limited activity with clearly established objectives and boundary conditions executed to gain knowledge, create a capability, or provide a service. The basic principles of project management do not change." In the document, NASA listed twelve key principles of project management. The first three of these twelve principles set the tone for the whole document and are good examples for emulation by other agencies:

1. *NASA welcomes and accepts the high challenge of carrying out projects*, expeditiously and at minimum cost, *that demand advances in technology [emphasis added].*

2. *While high risk is inherent in the work done by NASA, every effort will be made to* understand and quantify this risk before seeking Office of Management and Budget and Congressional commitment to a project. *The desire to obtain project approval or to proceed with project implementation should not be permitted to interfere with the adequate and comprehensive studies and cost analyses necessary to define the risk so as to minimize the potential for cost over-runs and schedule slips [emphasis added].*

3. *The Project Manager is the key individual charged with project execution. After project approval, he or she will have authority and responsibility for* all project decisions *and will make all recommendations to the management structure required for higher level decisions [emphasis added].*

Seven of the twelve principles used by NASA have to do with the project manager's selection, tenure, and duties. Some of the principles outlined in the NASA document are:

- The project manager will be appointed as *early in the project cycle as possible.*
- The project manager *will be expected to remain with the project until its completion.*
- The position of project manager is at such a level that it will be aspired to as a *career goal.*
- Incumbents of the project manager positions will normally *remain in them for extended periods.*
- Appointment of the project manager is made *only after in-depth discussions* with higher authority *during which they shall establish a clear understanding of the conditions under which the project is to be implemented and the resources that are to be provided.*
- The project manager's authority *follows a clear chain of delegation* that starts with the project approval document.
- The project manager, through the immediately surrounding institution, *must have direct control* of the systems engineering function and resources.
- It is essential that the project manager *develop control of the detailed technical content of the project together with associated costs, before* the *project or contract reaches* full momentum.

Dr. Hans J. Thamhain of Worcester Polytechnic Institute, in his book titled *Engineering Project Management,*[15] says that "managing engineering programs toward established performance, schedule, and cost targets requires more than just another plan. It requires the *total commitment* of the performing organization plus the involvement and help of the sponsor/customer community [emphasis added]." He then points out a proven, eight-step formula for success:

1. Work out a detailed project plan involving all key personnel.
2. Reach agreement on the plan among the project team members and their customer/sponsor.
3. Obtain commitment from the project team members and contractors.
4. Obtain commitment from management for the resources required to do the job.
5. Define measurable milestones, and track actual progress against these milestones.
6. Attract and hold high quality people.
7. Establish a controlling authority for each work package.
8. Detect problems early.

If problems are discovered, then the process goes back to step 1, where a new plan is devised, commitments to accomplish the revised plan are obtained, and the work continues. Thus, project management is really a closed-loop cycle in which definition is iterated through the correction of problems. This cycle requires a sensitive and responsible problem detection mechanism and a commitment to provide an immediate feedback into the planning and implementation process.

The reason that the project manager must be involved early in the process is that this is where the key commitments are made. The reason that continuity is so essential is that the project manager who made the commitments should be required to carry them out. It is likely that a new project manager may not feel obligated to carry out the commitments of a prior project manager. If a project manager knows that he or she will be held responsible for project commitments, it is more likely that much greater care will be taken in establishing these commitments. Because projects in the early phases are often in the sales or marketing mode, overoptimism is a hazard. A way of counteracting this is to assign an experienced project manager who really knows what it takes to do the job, or who has the authority to engage peer groups who do, and to hold him or her responsible for *doing* the project as well as defining it. Optimistic commitments made in the flurry of getting program approval are often difficult to meet when the facts of life are exposed under a hard-nosed, practical, realistic project manager. An experienced military project manager told us: "When a new project manager arrives at a program that is in trouble, he is either a whistle-blower in the first month—and there is an unwritten code against that—or he dives in and tries his best to bring things onto a proper course." Our conclusion is that the best of all worlds is to have the person who has to carry out the goals be the one to set them. The GAO[9] suggested that the ways to ensure a strong commitment to a project ranged from "*having experienced, confident project managers* [emphasis added]" to "increased involvement of the user and high-level prioritization of mission needs."

It is evident from the foregoing that there are three C's of vital importance in the project manager: Competence, Commitment, and *Continuity*. The first two are of little value without the third. The statement that is attributed to Abe Lincoln in 1864 while running for the Presidency is just as true now as it was then: "*It is not best to swap horses while crossing the river*!" The more popular way of stating this adage is "*Don't change horses in the middle of the stream!*" If we are to hold our project managers responsible for carrying out successful projects, we must give them the opportunity, time, and authority to do just that: to carry their projects through successfully. Strong action is needed to ensure that our project managers for major systems possess the special attributes needed to manage our major programs; that they are provided with a sustained career path in the project management field; and that they are committed, on a long-term basis, to their program and to its continuity.

Endnotes

1. Institute of Cost Analysis: *Proceedings of the First Annual ICA Symposium*, Institute of Cost Analysis, Washington, DC, 7 Dec 1982.
2. David I. Cleland and Harold Kerzner: *The Best Managed Projects*, Procurement Associates, Inc., Covina, CA, April 1986, GCS 8-86 D-1.
3. Lewis R. Ireland: *Project Management: Critical Success Factors and Keys to Effectiveness*, SWL, Inc., McLean, VA, August 1986, 58-07-86.
4. John R. Adams, C. Richard Bilbro, and Timothy C. Stockert: *An Organization Development Approach to Project Management*, Project Management Institute, Drexel Hill, PA, 1986.
5. Dina Rusor: *The Pentagon Underground*, Times Books, New York, 1984.
6. Benjamin S. Blanchard and Wolter J. Fabrycky: *Systems Engineering and Analysis*, Prentice-Hall, Englewood Cliffs, NJ, 1981.
7. General Accounting Office: *Hazards of Beginning Weapons Systems Production without Testing*, General Accounting Office, Washington, DC, June 1985, NSIAD-85-68.
8. General Accounting Office: *Weapon Performance: Operational Test and Evaluation Can Contribute More to Decisionmaking*, General Accounting Office, Washington, DC, December 1986, NSIAD-87-57.
9. General Accounting Office: *Strengthening the Capabilities of Key Acquisition Personnel (in DOD) Systems Acquisition*, General Accounting Office, Washington, DC, May 1986, NSIAD-86-45.
10. Gary Hart and William S. Lind: *America Can Win*, Adler & Adler, Bethesda, MD, 1986.
11. General Accounting Office: *DOD Acquisition Case Study of Army Tactical Missile System*, General Accounting Office, Washington, DC, July 1986, NSIAD-86-45S2.
12. General Accounting Office: *DOD Acquisition Case Study Air Force Small Intercontinental Ballistic Missile*, General Accounting Office, Washington, DC, July 1986, NSIAD-86-45S-16.

13. Department of Defense: *Guide to Resources and Sources of Information for Acquisition Research,* Department of Defense, Washington, DC, January 1980.
14. Project Management Institute: *Project Management Journal,* Project Management Institute, Drexel Hill, PA, August 1984, 0147-5363.
15. National Aeronautics and Space Administration: *Principles of Project Management,* National Aeronautics and Space Administration, Washington, DC, January 1982, NHB 7120.2.
16. Hans J. Thamhain: *Engineering Program Management,* Wiley, New York, 1984.

CHAPTER SIX

Private Industry and the Procurement Process

It is the policy of the U.S. government to acquire its major systems principally through contracts with private industry. Although government arsenals, research laboratories, and federally funded research and development centers still exist for the purposes of management, limited research and development, and prototype testing, the bulk of the work required to develop and produce our major systems is carried out by a private sector consisting of prime contractors, subcontractors, nonprofit foundations and corporations, universities, consultants, and suppliers of parts and services.

A key to technically successful new major systems and projects lies in the technology base, which is the source of new knowledge and techniques for future systems. Where does the technology base lie? In the government itself? No. It lies in the U.S. defense, construction, aerospace, and high-technology industrial base that has built up over the past several decades to develop our current military and civil systems. The dilemma of the government has been how to maintain a superb technological base without building up this base to a point where there is not enough money to support the industry. The U.S. defense, aerospace, and high-tech industries have been walking a federal budget tightrope that is already stretched almost to the breaking point. Each time an industry is added or starts growing, it takes a bigger rope to support it. The industrial base consists of the technological base and a production base. There are those who continue to push (unrealistically, we might add) for a huge military production base that might be required in the event of any size conflict from a limited war to a global encounter the size of World War II.

It is clear that the peacetime industrial base must be large enough to accommodate wartime mobilization. But it is also clear that overnight wartime conversion is not now possible. In any high-tech war of the future, there will be little time to tool up for production, as the "flash point" of hostilities would precede a conclusive victory in terms of months, weeks, days, or even hours instead of years or decades. The government has already responded to this knowledge through the buildup of stockpiles of weapons, fully recognizing that by the time production lines could be rolling, the war would be over. In the meantime, however, technological advancements must continue in order to be competitive in both peacetime and wartime.

Recognizing that we must continue to feed (and support with fruitful work) an already large industrial base, a new degree of self-restraint is needed to prevent the creation of more hungry industrial giants. Further, a portion of the industrial base we already have can be redirected to the important task of advancing technology so that future systems will be ready to accommodate ideas and advancements that have been previously reduced to practice. In the extensive interviews conducted by the Major Systems Acquisition Study Group in 1969–1970, industry leaders complained about being both informally and formally encouraged to add capital investments and to build up their capacities, only to be cut off and left holding the bag. Companies were led on by false hopes of winning big research, development, and production contracts to be paid for by money that did not exist in the federal budget. More recently, the Competition in Contracting Act (which is discussed in detail later) tended to raise false hopes among companies that were not fully qualified to participate in major systems, projects, or their supporting services and parts contracts. This overreaction to the intent of the Competition in Contracting Act was recognized by the Congress in its fiscal year 1987 Authorization Act where it called attention to the insufficient attention being given to the *quality* of goods and services. The Act was causing more companies to bid on government work; but were these companies qualified?

This environment leads to a need for adoption of some important management philosophies and a new procurement environment to prevent disillusionment among those who could invest huge sums in anticipating government work. First, the dynamic national planning process must include a consideration of existing and available, as well as new and growing, industrial capabilities in structuring the long-range major systems acquisition plan. Officials must avoid leading industry on by false promises that cannot be supported by the budget. Affirmative action is needed to distribute workload to more fully use industry capacity. Instead of selection pressures for major systems and projects being sensitive to geoeconomic factors solely on a case-by-case basis, as they are now, the effect on the entire nation's technological and industrial base must also be considered in the context of the total systems and projects planned for the next score of years. But because of the probable rapid progression of potential major conflicts, a production mobilization base makes sense only for short lead time items. Other measures, such as those proposed by Jacques Gansler in *The Defense Industry* to direct foreign military sales into the mobilization base in the event of a national emergency and to integrate civilian and military capabilities in heretofore defense-only plants, are ways of providing a defense industrial mobilization base without excessive buildup of capabilities that drive the systems acquisition and selection process. Our view is that the focus should be principally on the job to be done rather than on the industry to be kept alive, as Gansler advocates.

Affirmative programs for retraining and retooling of industry, separation of the research and development base from the production base, and a less regulated, highly motivated industry responding to inherent rather than artificially induced competition will be strong forces that will bring about the adaptations and adjustments needed by industry for future major systems and projects. The ability to respond rapidly lies in industry rather than in government; but the government must have efficient and flexible tools with which to activate, motivate, control, and manage. The principal tool that makes it possible to use private enterprise in these large undertakings is the government contract, and the process of engaging private industry for this work is called the procurement process. This chapter will show that phased, selective deregulation and simplification of the procurement process will help create an environment in which industry can respond rapidly to national major systems needs and confidently make long-term commitments.

The procurement process is guided by a mammoth document called the FAR (Federal Acquisition Regulation) which contains fifty-three parts in two volumes and eight major subchapters ranging in subject from "Competition and Acquisition Planning" to "Clauses and Forms." Over 1,000 pages long, the FAR also has supplements for each agency which contain from a few to 1,000 pages each. Although great efforts have been expended to try to reduce the size of this regulation, it has continued to grow both in magnitude and complexity. The FAR is continually updated as new government procurement laws and policies are introduced in Congress. (At last count, these laws numbered *over 4,000,* and more are being enacted in each session of Congress, some as stand-alone laws, but many that are attached to other laws or to authorization or appropriation bills (see Chapter 7). Further, a whole body of prescriptions exists in case law that has resulted from Comptroller General and Board of Contract Appeals decisions. Although some of the laws and their resulting regulations apply to services, parts, and supplies, the bulk have a direct or indirect effect on our government's major systems and projects. The introduction and seemingly constant changing of these laws and regulations has resulted in an enormously complex acquisition process. As a result, there is a growing need for acquisition simplification and reform. Some steps along this line are, in fact, being taken by the Department of Defense and other agencies. But, as we will show later, some of these "reforms" are merely unproven reorganizations or laws that are adding to the complexity and time required for government procurement rather than simplifying it. Procurement professionals and reformers alike have tried to be innovative and creative in developing whole new disciplines that would supposedly correct all of the problems in buying the big systems. Some have been successful, but many have ignored the proven and workable fundamental principles of planning, design, development, management, and financial control. These quick-fix reforms have failed to recognize that phased, selective deregulation accompanied by continuous moves toward simplification will require the same type of meticulous planning, timing, and stepwise implementation that we are advocating for major systems themselves. Failure to use an orderly, systematic, time-phased approach to deregulation has already had its ill effects in the farming, communications, and airline industries. Clearly, a well-planned transition with selective rather than wholesale deregulation is needed.

Perhaps the second greatest critic (second only to the media) of government procurement is the U.S. industry that provides major systems and their parts, support, training, operations, and logistics to the government. The government-contractor community largely feels that it is over-regulated, over-controlled, and over-audited. Murray L. Weidenbaum, director of the Center for the Study of American Business, Washington University, St. Louis, Missouri, writing in the "Christian Science Monitor" as reported in "Aerospace" magazine[2] said:

> *The existing detailed bureaucratic oversight of military production is counterproductive. It has not prevented numerous cost overruns and problems with quality control. The record of competition in the private marketplace, in contrast, has been impressive over the years—driving down costs and providing efficient production.* Thus, substituting economic incentives for detailed surveillance of the operations of government contractors is likely to produce more defense output for the dollar. . . . [emphasis added]

Norman R. Augustine, president of Martin Marietta Denver Aerospace,[3] plotted the growth of government regulation against that of a weed and found a striking correlation. He said, "Those of us in industry certainly are worried about the amount of regulation we see, even at a time when there is allegedly a movement toward deregulation."

From an industry viewpoint, the obstacle course to achieving prominence in government contracting is tough, but the rewards are great enough to counterbalance the hardship it takes to get there—at least so far. Despite efforts by the Packard Commission[4] and others to correlate the government systems procurement process with the commercial marketplace, little headway has been made in adopting private sector methods of achieving efficiency, economy, and effectiveness.

The fundamental problem with our procurement laws, incentives, procedures, and policies is that we have tried to force both government agencies and contractors into sound management practices and successful performance through contract clauses and legal incentives rather than through adoption of proven-to-be-successful planning, engineering, and management approaches. The procurement community has done the very best it could with the complex system we the people have permitted. But experience has shown that "innovative" contracting is no substitute for sound engineering and management. Much has been done by the procurement community in recent years to attempt to broaden the responsibilities, scope, and viewpoint of procurement officers, to include functions supplemental to the simple act of preparing and negotiating contracts. This has been evidenced by a trend in titles of the process from *contracting* to *procurement* to *acquisition*. Procurement was originally intended to encompass more than contracting, and acquisition was intended to include more than either. But little has really been accomplished toward providing a broader responsibility for contracting personnel, principally because their only implementing tool is still the government contract, and the government contract is highly regulated and controlled through legislation. Name changes have not altered the principal function of procurement officers or more directly addressed the three *e*'s (efficiency, economy, and effectiveness) in major system acquisition. We propose that procurement officers be given greater latitude to make decisions by reducing and simplifying controlling laws, policies, and procedures.

The Extent of Competitive Optimism

In 1962, the Graduate School of Business Administration of Harvard University published the results of a classic study by Merton J. Peck and Frederic M. Scherer, titled *The Weapons Acquisition Process: An Economic Analysis*.[5] In the report of this study, Peck and Scherer put their finger precisely on a major problem that existed in the government procurement of major systems then, and still exists today. In a section titled "The Extent of Competitive Optimism," they tell precisely why industry continues to underbid on government contracts. They demonstrate the reasons, motivations, and probable effects of competitive optimism on the government "nonmarketplace."

As Peck and Scherer point out, competitive optimism often results in understating costs, establishing unrealistically short-time schedules, and specifying unrealistically high-performance goals. Worse, the government tends to encourage underestimates by imposing unrealistic budget limitations while insisting on higher and higher performance. As long as procurement time-phasing and contracting methods continue to foster these practices, the chronic problems of missed milestones, cost growth, and performance shortfalls will continue.

The cost reimbursement contract is a principal tool in the conceptual studies, development, and perfection of our major systems. Fixed-price contracts are used where the job has been thoroughly defined. Fee or profit is proportional to the overall size of the job but is based on risk, investment, and type of contract. Long-run "adequate" profits are a part of the objectives of government contractors just as they are of commercial contractors, but they are not the only motivators. This has been documented thoroughly in government funded studies sponsored by the Air Force, Navy, and other government agencies[6,7] conducted by the School of Business at the University of Notre Dame. These studies involved extensive questionnaire programs, literature research, workshops, in-house studies, and reviews of previous government, academic, and industry studies. A principal finding of the Notre Dame study was that *"THE CONTRACT TYPE IS NOT THE DETERMINING VARIABLE AS TO THE OUTCOMES OF CONTRACTS"*[6] If the contract type, the fee arrangement, and even the after-tax profits are not the only motivations of government contractors in achieving excellence, as shown in the study, then what *are* the major motivating factors?

The many motivating factors listed in the government funded studies can be condensed to ten: (1) corporate survival, (2) reduced risk, (3) volume of business, (4) corporate growth, (5) expanded market potential, (6) reputation, (7) adequate long range profits, (8) tax incentives, (9) executive incentives, and (10) other socioeconomic objectives. Notice that only one of the ten is associated with profits. Corporate survival is the number one concern of government contractors, as it is with just about any business. Survival includes the need to maintain an adequate cash flow to pay its employees, to cover its overhead, and to maintain a base for potential future work and for potential growth.

Phillip P. Oppendahl, a Commander in the U.S. Navy, while a student in the Defense Weapons Management College, produced a 1977 research project titled "Understanding Contractor Motivation and Contract Incentives." He summarized the work of ten other previous researchers and concluded that *incentive-fee contracts did not work.* The motivator was not solely profit, and the type of contract was not the major variable in contractor behavior. He developed a list of motivators that applied during the various stages of the growth of a company which started with "survival" and ended with "prestige." In order to ensure survival, the government contractor is searching for ways to reduce *risk*! This can take the form of cost-reimbursement contracts, inflation clauses, guaranteed follow-on buys, and improved definition of the work to be performed. Recently, profit policy for government contractors has been changed and is no longer based on costs. Allowable profit is now based on risk, type of contract, and corporate investment. Procurement officials structuring fee based on risk must be careful not to actually *encourage* excessive risk-taking. It is to the benefit of *both* parties that risk-*reduction* techniques such as systems engineering, analysis, simulation, and testing be used extensively to *eliminate or at least reduce risk* early as well as throughout the project.

Allowable overhead costs under government contracts include bid and proposal costs as well as independent research and development reimbursements. Higher volumes of business produce greater overhead coverage and allowances and enhance the company's position for follow-on work because of a greater capacity as well as an improved competitive position in new technology. Current capacity can often be a major influencing factor on winning new government contracts because there is a tendency for source selection officials to favor firms that can demonstrate under-utilization of a previously-acquired or developed technological facility or personnel base.

A major industry motivator to stay in business at the highest gross sales level possible is the recovery of overhead costs. Non-government contractors and the general public are probably unaware of the overhead allowances available in government contracting and the cost elements covered by this overhead. In a study we conducted of nine large, high-tech aerospace firms, we found that engineering overhead ranged from 75 percent to 127 percent with an average of about 100 percent. This means that, for every allowable direct engineering dollar spent, another dollar is allocated to the engineering overhead pool. Thus, a contract containing $100 million of direct engineering labor dollars yields another $100 million in overhead. This fund is used partly to pay labor burden costs, some of which are required by government regulations and some of which are established by the company. These labor burden costs include the company's contribution to social security, retirement plans, and health and life insurance, as well as paid holidays, paid vacations, and bonuses. By far the largest part of overhead costs are used for housekeeping and other costs of doing business that cannot be easily allocated to specific projects. Housekeeping costs include communications, custodial, heating and cooling, power, water, lighting, insurance, maintenance, operating supplies, rental of buildings, and waste disposal. Other costs that are usually permitted as legitimate overhead charges in government contracts are bid and proposal costs, independent research and development, amortization, claims, depreciation, industrial relations, environmental protection, and material handling, storage, preservation, and inventory control costs.

Manufacturing overheads for the large, high-tech aerospace companies are usually about 50 percent greater than engineering overheads. In our study we found that manufacturing overheads ranged from 148 percent to 202 percent of direct manufacturing labor dollars, with the average hovering around 160 percent. Thus, for a direct manufacturing cost of $100 million, another $160 million is added to the manufacturing overhead pool. The higher amount for manufacturing overhead is caused by the greater use of power, space, maintenance, supplies, and investment for large, high-tech manufacturing equipment.

In addition to engineering and manufacturing overhead, government contractors are allowed an additional percentage of all direct plus labor overhead costs to cover "general and administrative" (G&A) costs. These costs are usually, but not always, used to support home-office administrative expenses for divisions of companies that have corporate headquarters elsewhere. A list of allowable G&A expenses includes administration, advanced design, advertising, corporate expenses, executive salaries, financing, marketing, personnel administration, research, training, and corporate taxes. Although the cost elements included in G&A expenses vary significantly from company to company, these elements are usually paid for by either overhead, G&A, or the fee paid under government contracts. Allowable G&A expense percentages of total costs for the nine companies ranged from 10.4 percent to 25.4 percent and averaged 20 percent. In addition to all this, government contractors are awarded a fee of up to 12 percent and which now averages about 9.5 percent. Assuming a 40/60 split of engineering and manufacturing, one can calculate that a government contractor can gross an average of up to $1.75 in addition to every direct dollar spent on producing a major system or project. The direct dollars are those that pay for the labor and materials to do the job, and the other added dollars are necessary to keep the company in business, to provide a place to do the work, and to support the employees.

The financial vice president of one large, highly successful aerospace defense firm once told us that a young Air Force lieutenant, just appointed to head program control on an Air Force project, didn't like the idea of paying overhead costs on the project. The lieutenant agreed that the direct labor and material costs were legitimate, but that his project should not have to share overhead costs with all other projects. The executive (the now-retired Ed Dohl of TRW) had to "sit the lieutenant down" and explain why overhead is needed and why its elimination would result in the work being performed out in the street with no telephones, restrooms, heat, cooling, light, and, in fact, with no company at all to support it. Government contractors need these add-on costs to stay in business, and, with bigger contracts and larger overall business volume, they have a greater opportunity to continue to survive and even to grow because their overhead and administrative pools are healthier.

Corporate growth and expanded market potential have been cited in many studies as two other strong motivators of government contractors. Many firms would rather increase their gross sales than make a higher percentage profit. This goes back to a fundamental law of survival that a company that is growing and continuing to explore new markets can continue to pay and to reward its executives and employees, while one that is gradually going out of business can continue to do neither very well. Employee ownership in large government contractors has not yet reached a level where corporate profits take precedence over the individual paycheck. Hence, everyone in the company is looking for more business, bigger contracts, and a guarantee of follow-on work. Stockholders benefit, of course, from increased volume of business and increasing net worth of the company.

A study by the Logistics Management Institute (LMI) into the foundations of incentive theory concluded that "extra-contractual factors dominated and directed behavior." Company growth, share of the market, carry-over benefits, and follow-on business were listed as major motivators. As early as 1963, Mr. Gordon Author, then head of pricing for the Air Force, in a New York meeting of the American Management Association, listed the potential for follow-on production and the utilization of current capacity as two major factors that had an impact on corporate behavior. Even earlier, in 1959, Congressional hearings held by Rep. Carl Vinson revealed that although profits were modest, the return on net worth for aerospace contractors average "an astonishing 71.3 percent."

Despite the move toward bigger contracts, higher gross sales, and greater follow-on potential as major motivators, government contractors are also concerned about their reputation. In the numerous studies of contractor motivators, factors such as establishing and maintaining technical excellence, attracting highly qualified and well-known analysts and executives, maintaining a good public image, and establishing organizational prestige were listed as significant drivers of corporate behavior. Contractors realize that a poor reputation in cost estimating and cost control can have a damaging effect on their ability to acquire the very follow-on business that they covet. They are finding that merely maintaining or capturing a larger share of the market may not be sufficient. Maintaining credibility with the Congress and the public in regard to long-term quality, excellence, and competence in helping to serve the public need is also an important corporate goal. Ethics initiatives being taken by industry itself and by industry groups argue that a gradual, phased deregulation of the procurement process may now be timely.

Of lesser but still significant importance to industry are adequate *long-term* profits, tax incentives, executive incentives, and "other socioeconomic objectives." The "other socioeconomic objectives" motivator has an obscure but powerful influence on the ability of a major government contractor to survive and to grow. These include the use of an available work force; the impact on a local economy; the ability to use minority, handicapped, or otherwise disadvantaged personnel and subcontractors; and philanthropic endeavors that help those in need in this country as well as in others. Large government contractors are increasingly viewed as not only holding a public trust to perform well in producing systems and projects, but as holding a social obligation to play a significant and helpful role in the social and economic welfare of cities, states, regions, and countries in which they are located. It is the responsibility of the Congress and the executive branch agencies to recognize that legislation and directives may no longer be needed in some socioeconomic areas because the goals of earlier policies are already achieved, and that many companies have now changed their way of doing business from that used when these laws were first put into effect.

Since the major motivators of government contractors have been identified in detail, documented, ordered in priority, and quantified, these motivators should be used to encourage excellence not only in product adequacy but also in estimating and scheduling realism, risk reduction, productivity improvements, cost effectiveness, and management control of large projects and systems. Natural incentives should be retained which play on these basic contractor drives and needs rather than focusing on contractual innovations which do no more than address the symptoms of more deep-seated problems.

We propose a management philosophy that includes positive incentives that address these underlying motivations. Excellence must be rewarded for a company to survive and grow. Incentives for excellence can take many forms. Companies that have demonstrated excellence in setting and meeting goals could be rewarded, for example, by being given preference on bidders lists. Firms that have demonstrated efficiency, economy, and effectiveness in providing the government's needs could be rewarded by giving them greater assurance of corporate survival, reduced risks, a higher volume of business, and a greater potential for corporate growth in the government marketplace. All of these are major motivators. The potential rewards can be synchronized with them.

Intangible rewards to enhance prestige and company reputation could be provided in recognition of superior performance. During World War II, outstanding government contractors were presented the "E" for Efficiency Award by each of the major services. This type of award could be reinstated, publicized, and presented each year to the contractors that have demonstrated high productivity, technical excellence, and an ability to meet their commitments. We believe that emphasis on elimination of negative factors, such as fraud, waste, and abuse, must be matched by emphasis on positive incentives to encourage efficiency, economy, and effectiveness. Management philosophies and programs must be instituted that will reward good performance, instill pride of good workmanship, and elevate the image and prestige of those contractors who do good work. Auditors, who traditionally look for improper or illegal actions by government contractors, should be trained to look for good and beneficial actions as well. These good and beneficial actions can be recognized, rewarded, and promoted in future projects. Top executives of major government contractors who display excellence and who have led their firms to outstanding performance in the public's best interest should be recognized and rewarded by the President and Congress.

The High Cost of a Little More

Along with incentives for outstanding performance, there is a need for positive incentives to keep the costs of major systems within the bounds of reason. Norman Augustine[3] points out several reasons for high costs, the most significant of which is the enormous cost of getting that last ounce of performance. He plotted the cost versus performance of several well-known products and services to prove his point. For example, he considered the cost of 35 mm optical lenses with increasing focal length, the baseball player's salary as batting average goes up, an airplane's cost with increasing mach number (speed), a diamond's cost with grade or quality increases, and machined parts cost as tolerances increase. In each case, the last available small improvement in performance is enormously expensive. Kirkpatrick and Pugh[3] also demonstrated that a country simply cannot afford, even within the limits of the gross national product, to obtain the last ounce of performance for all major systems. Both call for a halt to ridiculously high specifications and for a broader look at overall "force-structure" effectiveness or team effectiveness versus cost, rather than squeezing that last ounce of performance out of each individual system. Evolution toward a rational, thrifty budgeting and planning process that is based on long-term national goals and objectives employing design-to-cost principles to stay within budget guidelines will result in developing excellent overall results. When companies must bid with the knowledge that meeting cost and schedule goals is important to future corporate health, they will be strongly motivated to estimate realistically and then to stay within estimated costs. The effort to be realistic in setting performance goals to begin with, however, is a joint responsibility of industry and government.

Admiral Rickover used to say that, when the temptation surfaces to go after that last enormously expensive ounce of performance, what is needed is a "cutter-offer." With the modern tools of cost analysis, it is possible to identify ahead of time where the cost starts rising inordinately with increasing performance, and to provide a "cutter-offer" to stop the continued growth. This will increase the demand for courageous decisionmakers who will make that difficult but necessary decision to cut off performance specifications (speed, attitude, range, firepower, and so on) at an economically achievable point and to save the remaining performance increments to be achieved by the next generation system. Industry has the responsibility to identify where these decision points are, and the government project manager has the responsibility to make the decision at the most economical break point and to make it stick.

Competing for Competition

Competition, like motherhood and apple pie, is good. Certainly in this free, democratic society, where innovators, entrepreneurs, and inventors are encouraged to come up with new and improved products and services, competition should be encouraged. However, should competition be legislated, and does legislated competition foster cost consciousness, savings, and excellence? If competition is good, how does it fit into the government procurement process in general and into the major systems acquisition process in specific? Finally, does *induced* competition address the underlying motivations of government contractors or does it merely treat the symptoms? And why does the government want to foster competition in its acquisitions?[8–10]

To answer these questions, first it is necessary to define the noun, *competition*, versus the verb, *to compete*. According to the 1983 edition of the *Random House Unabridged Dictionary*, competition is "the act of struggling or rivalry for supremacy." To compete means "to meet, coincide, be fitting, or be capable." The definition goes on to say that "Compete implies having a sense of rivalry and of striving to do one's best as well as to outdo another."

A still valid, moderate, common-sense approach to encouraging competition and a phased-procurement strategy was expressed in the Commission on Government Procurement's Major Systems Acquisition Study Group's report[11] as follows:

> *A fundamental purpose of rewarding commercial participants in the system acquisition process is to provide the wherewithal for maintaining a healthy and efficient industry.* Neither technical nor price competition can work their effective free market purposes in a situation that does not reward enterprise and efficiency *[emphasis added].*
>
> In the main, the adoption of an incremental acquisition policy would tend to reduce risk and to create a higher assurance of a fair return on enterprise at various stages in the acquisition process itself *[emphasis added]. Basically a fair return should be paid for satisfaction of the central specifications for a system program with substantial bonuses or penalties being reserved respectively for superior performance or through inferior performance.*
>
> Obviously, competition for a production contract would be, as with other aspects of a highly flexible acquisition policy, highly dependent on circumstances. The possibility that specialized design teams might be created and maintained independent of any single production facility represents one such circumstance. Others can be conceived. In the end, however, the principal criterion for the maintenance of competition past any particular stage in the major system acquisition process should be the advisability of eliminating—at that point—a particular design approach or a particular source. *[emphasis added].*

Ten years after the Commission on Government Procurement issued its final report, Presidential Executive Order 12352 was issued as part of President Reagan's "Reform 88" Program. This order stated that "to make procurement more effective in support of mission accomplishment, the heads of executive agencies engaged in the procurement of products and services from the private sector" would have to take certain actions, one of which follows:

> *(d)* ESTABLISH CRITERIA FOR ENHANCING EFFECTIVE COMPETITION AND LIMITING NONCOMPETITIVE ACTIONS. *These criteria shall seek to improve competition by such actions as eliminating unnecessary government specifications and simplifying those that must be retained, expanding the purchase of available commercial goods and services, and, where practical, using functionally-oriented specifications or otherwise describing government needs so as to permit greater latitude for private sector response. . ."*

Note in the above that deregulation and simplification of specifications, combined with reduction of specificity, must accompany the enhancement of competition. Also note, however, that the order referred to "goods and services" and not to major systems. Where commercial goods and services are not available, such as is the case for parts and specialized services in major military weapons, *more* definition of the job, rather than less, is required.

An "Executive Committee on Federal Procurement Reform" was established to implement this Presidential order, and a seventeen-person task group developed, among other documents, a thirty-seven page document called "Module B: Criteria for Enhancing Competition," which was recommended to the heads of all executive departments and agencies in a memorandum from the Office of Management and Budget in February 1984.[8] Shortly thereafter, Congress enacted the Competition in Contracting Act of 1984[12] and other laws, one of which required the head of each agency to submit a report to Congress each year which would include:

1. A specific description of all actions that the head of the executive agency intends to take during the current fiscal year to
   a. increase competition for contracts with the executive agency on the basis of cost and other significant factors; and
   b. reduce the number and dollar value of noncompetitive contracts entered into by the executive agency; and
2. A summary of the activities and accomplishments of the advocate for competition of the executive agency during the preceding fiscal year.

The act did not require simultaneous reporting of the reduction of the numbers of controls and regulations placed on government contractors, nor did it state that it only applied to goods and services and not to major systems and projects.

The effect of these policy directives and laws is that procurement officials in the federal departments, agencies, and services are now attempting to produce the largest number and the largest dollar value of competitive contracts, while what really should count is the total dollar and time savings both in industry and government in procuring its systems, projects, goods, and services. In the process of using applicable but not crucial criteria for success, the emphasis is placed on related but less important goals, and a new discipline or procedure has been elevated to an inordinately high degree of importance.

In July 1986 a magazine named *Military Business Review,* subtitled *Advocating Competition in the Marketplace,*[9] was published. On the cover of the magazine appeared two brigadier generals and one rear admiral who had been selected as "Competition Advocates" by their three services, the Army, Air Force, and Navy respectively. Inside were charts that showed that the percentage of dollars competed by the Army increased from 40.3 percent in fiscal year 1982 to 46.9 percent in 1985; the Navy's increased from 27 percent in FY1982 to 45 percent in 1985; and so on. This document, and others that have been published since then, showed that there is definitely a race to compete larger percentages of contracts and larger percentages of the procurement dollar. But this effort has not resulted in an overall decrease in the costs to the government of major systems and other goods and services.

Recent studies have shown that there is a sizeable penalty for artificially stimulating competition. Competition is a natural phenomenon that occurs in any free society, and it tends, in the long run, to produce better, more timely, and more economical products and services. In some instances the introduction of competition into the government procurement process has indeed reduced the contracted costs for goods and services that support major systems as well as the contracted costs for the systems themselves. But in our interviews with government and industry procurement and project management personnel and in many recent publications, we found that the competition in contracting policies now being followed contributed to *increased procurement lead times, increased procurement administration costs, and degradation of quality in the end product*. In the major systems arena, forcing competition means that more than one source must be selected or qualified for one or more of the procurement phases. Starting out with more than one potential source and keeping it alive to encourage efficiency in the other, as was the case in the MLRS project, is a germane approach; but reopening competition in the later stages of the procurement cycle (full scale development, production, or operations) requires educating or re-educating a new or previous supplier and transferring of data, often considered proprietary by the incumbent, to a company's competitors. Increased administrative costs and longer time delays within both industry and government are caused by the requirement to force "full and open competition."

According to Earle C. Williams' article in the December 1986 *Contract Management*[13] "The Competition in Contracting Act and other 'innovative business strategies' are going to seriously damage the nation's defense efforts if they are not reversed." Williams claims that there are those who are trying to "pervert the concept of full and open competition by attempting to restrict competition to the one element of price." He points out that there are occasions where price competition is *not* best for the government, and that Congress and the executive branch must shift their emphasis from "competition scorecards" to "competition for quality."

Mr. Williams' article, as well as inputs from a number of other sources, tell us that the average time from initial discussions to contract award is now eighteen to twenty-four months for a major system or a research and development project. This represents a 30 to 50 percent increase in time for the average procurement cycle. With technology moving rapidly and continuing to accelerate, this increasing procurement lead time could cause the final product to be one-and-a-half to two years behind the current state-of-the-art by the time it is delivered to the user. According to Williams, a company's past experience and proven track record for excellence in service or product quality are being given less weight in evaluations, with the major emphasis being on "clever proposals and the bottom-line price."

Frederick D. Biery of the Analytic Sciences Corporation presented a paper at the 1985 National Conference of the National Estimating Society where he provided an example of cost reductions in TOW missiles resulting from the introduction of competition. He said "the results of other cost-benefit studies also show, however, a wide range of outcomes. *Thus it becomes imperative that competition be assessed on a case-by-case basis*" [emphasis added]. Michael N. Beltramo, in an article in "The Journal of Cost Analysis" in the fall of 1986 summed up his findings and thoughts about competition by saying, "It should not be applied as a panacea nor as a substitute for sound management practices."

Competition is good if it is the type that results in greater overall cost effectiveness in acquiring a needed system or project. Too much emphasis on competition for its own sake, however, can truly become counterproductive. Competing for excellence must include competing for management excellence and technical excellence as well as low cost. Management and technical excellence include an ability to establish and meet commitments. We, the taxpayers, are looking for reasonable final costs at the project's completion, not just lower promised costs at its inception. Artificially induced competition tends to bring about the latter, not the former. A vital part of any phased deregulation and simplification process would be to return to a common-sense approach to source solicitations and selections that results in the best people being selected to do the work, rather than having the most people respond to bid requests.

Removing Socioeconomic Objectives from the Arena of Government Procurement

The Commission on Government Procurement made a valid recommendation that should be a part of any far-reaching procurement reform: "to reexamine the use of government procurement as a socioeconomic vehicle." Socioeconomic objectives are the laws that require government prime and subcontractors to buy American goods; to employ qualified persons regardless of race, creed, color, religion, sex, or age; to employ the handicapped; to conserve energy; to use small businesses; to protect the environment; to employ economically disadvantaged persons; to use labor in labor surplus areas; and so forth. The recommendation was explained by the General Accounting Office as follows:[14]

> *The leverage of Government procurement has been increasingly used to pursue social and economic ends, such as fair employment practices, small and minority business participation, and rehabilitation of handicapped persons.*
>
> *The Commission did not question the merit of these programs. Rather, to avoid impairing the Federal procurement process, the Commission stressed the need to reexamine existing applications, streamline administration, and consider other more direct means (alternatives) to implement new socioeconomic programs.*

Although it is difficult to determine the exact cost impact of socioeconomic objectives, it is generally conceded in industry that they create enormous costs for government prime contractors, subcontractors, and suppliers of parts, raw materials, and services. The high costs are brought about by a requirement for extensive reporting of progress and requirements that reflect on the ability to purchase the most cost-effective parts, materials or services. The commercial marketplace, which is often compared with the government as an ideal target or objective for efficiency, is not forced by procurement law to adhere to contractually imposed socioeconomic objectives. The proper place for these laws is in the civil judicial system and not in the government procurement process. In fact, many states and local governments are taking the initiative to enact these laws and institute these practices.

In a report to the Congress five years after the Procurement Commission submitted its final report,[14] the GAO stated that the executive branch "has accepted the recommendation but sees little receptivity in the Congress."

The time has come for Congress to review the socioeconomic laws that are on the books to determine if they can be consolidated or reduced, and to determine if there are other means of encouraging social, environmental, and political changes than by placing these burdens on the federal procurement process.

Reversing the Trend Toward Increased Regulation

At a time when other industries are being deregulated to enhance cost reduction by fostering increased competition in the open marketplace, the industries that provide major systems and related products and services to the government are becoming more regulated through laws such as the False Claims Amendments Act (Public Law 99-562), the Program Fraud

Remedies Act (Public Law 99-509), the Anti-Kickback Enforcement Act (Public Law 99-634), the Warranty/Guarantee Requirements of the Defense Procurement Reform Act of 1984 (Public Law 98-525), and by many provisions in the authorizations and appropriations acts that Congress enacts each fiscal year. Many other such laws are being proposed every year by the Congress. Examples are a Federal Contractor's Self-Governance Act, a Financial Fraud Detection and Disclosure Act, and "Omnibus Procurement Reform Amendments" to Authorizations. The omnibus amendments[15] include such subjects as:

- Conflict of interest in Defense procurement (slowing down or stopping the revolving door);[a]
- A provision to preserve independent "should-cost" analysis for major systems contractors;
- A provision to require work measurement analysis for government contractors
- Protection of whistle-blowers
- Limitation on the use of undefinitized contracts
- Certification of allowable indirect costs;
- Competitive prototype strategy requirement
- Release of technical data.

None of these amendments are harmful themselves, but the attempt to solve problems on a piecemeal basis has created a maze of requirements that are becoming increasingly difficult for procurement officers and industrial participants to sort out and responsibly implement.

The desire to keep government contractors honest dates as far back as the Revolutionary War, when George Washington suggested hanging wartime speculators as an example to others and said, "No punishment in my opinion is too great for the man who can build his reputation for greatness upon his country's ruin." Congress enacted legislation during the Civil War, some of which still survives, relating to bribery and corruption in government contracting, but it did nothing to regulate profiteering. For example, in one case, the Secretary of the Navy appointed his brother-in-law as an agent to purchase ships at a 2.5 percent commission when the going rate for ship purchasing agents was 1 percent. Later, procurement laws with meaningful and enforceable penalties were put into place that eliminated the opportunity for fraud, but over the years both the severity and enforcement of these penalties has decreased, and the complexity of the laws themselves has reduced their impact.

Ever since, Congress and the executive branch have been trying to legis-

[a] The "revolving door" refers to the industry practice of hiring government officials who were influential in awarding contracts to the industrial firm that hired them.

late honesty in government procurement by using various forms of contracts, audits, and purchasing procedures. Not that we are against honesty; but the record shows that it cannot be legislated without means of detection, enforceability, or penalties that are truly deterrent. The grandfather of morality legislation was the "Truth in Negotiations" Law (Public Law 87-653) enacted in 1962 which said:

> *A prime contractor or any subcontractor shall be required to submit cost or pricing data . . . and shall be required to* certify that to the best of his knowledge and belief, the cost or pricing data he submitted was accurate, complete, and current *[emphasis added].*

Despite these and many other laws enacted since, evidences of fraud, waste, and abuse are continuing to surface. According to the Defense Contract Audit Agency (DCAA), 313 instances of apparent fraud were reported for the first eleven months of fiscal year 1986, as compared to 199 in fiscal year 1985 and 126 in fiscal year 1984. Fred Newton, DCAA Deputy Director, states that they are now finding more sophisticated kinds of fraud and finding it in less obvious places.[16]

In the meantime, some rather interesting legislation has been introduced in Congress. Amendments being considered to the Anti-Kickback Act of 1946, *many of which reflect input by industry representatives*, include the following provisions[16]

- Distinguish between knowing and unknowing violations of the act. The government would receive twice the amount of each kickback plus up to $10,000 from persons who "knowingly violated the act." In any other case, violators would be responsible only for the amount of each kickback.
- State that evidence that a contractor reported possible violations should be considered as a favorable indication of a contractor's responsibility in a suspension or debarment proceeding.
- Specify that a prime contractor is responsible for following *reasonable procedures* to prevent violations of the act in its own operation.
- Remove from the bill a requirement that each contractor employee make an annual declaration of all compensation received from subcontractors.
- Eliminate a provision which authorized the Government to terminate prime contracts affected by subcontractor kickbacks.

The wording of these provisions gives the impression that the people who drafted them want to avoid any responsibility or penalty for wrongdoing either in their own organizations or in ones under their control. Laws that encourage avoiding the consequences of wrongdoing, rather than accepting responsibility to make appropriate corrections, just further speed up the merry-go-round of regulatory provisions that must be enforced by proof of *intent*, which is always very difficult to obtain.

The Federal Contractor's Self-Governance Act would require federal contractors to establish and maintain a system of internal accounting and administrative controls sufficient to ensure that "long-term cost and pricing data submitted or made available to the government is accurate, complete, current, and in conformance with government regulations." The bill would also require contractors to report on its systems to the Securities and Exchange Commission and would require each contractor covered by the bill to "have an independent public accountant review the system to determine whether its internal controls are sufficient, test transactions under each major government contract, and prepare a report on the findings for the company's annual report." This law would really accomplish no more than the Truth in Negotiations Law, but it starts to tell the contractor *how* to certify that his costs are "accurate, complete, and current." Until the Federal government itself is willing and able to have itself reviewed by an independent public accountant and impose the results of the review on itself, it will be difficult to impose further reviews on contractors, each of which already has its own certified accounting firm. Here again, we are attempting to legislate something that should come naturally. Legislating "self-governance" seems to take away from rather than add to the incentives for responsible practices. Natural self-governance, coupled with effective motivational management philosophies and incentives, would clearly put the responsibility on industry to provide the high quality products needed for the public good. Enforceable laws that prevent fraud and abuse are still needed, but they can be balanced by management initiatives that identify and reward excellence. Again, we are not against morality, only against the fruitless attempts to legislate it through largely unenforceable laws. The companies that have proven they can be trusted to perform prudently and in the public's best interest should be allowed to do so without excessive controls.

The direct expense of fraud, waste, and abuse programs to the government itself is immense. In the Air Force alone, which is merely one branch of one department of the federal government, and one which procures a large number of major systems, $175 million was spent in fiscal year 1985 to conduct audits, investigations, and inspections. There were 1,894 audits, 1,562 inspections, and 1,722 Air Force Office of Special Investigations (AFOSI)-conducted fraud-related investigations completed that year alone according to the FY'87 Air Force Acquisition Statement.[17] These investigative efforts resulted in over 1,800 cases referred for prosecution by Air Force officials and another 200 turned over to federal and local investigative agencies.

A President's Council on Integrity and Efficiency (PCIE) was created in 1981 as a part of the administration's program to "eliminate fraud, waste, and abuse." After four and one half years of operation, the PCIE reported $63 billion in improved use of funds, 14,300 successful criminal prosecutions, recoveries from investigations in excess of $500 million, and over 14,000 "administrative sanctions."[18] The above facts tell us two things. First, despite all of the legislation enacted since George Washington's time, fraud, waste, and abuse are still with us, and second, either the occurrences are increasing or the method of uncovering the occurrences are vastly improved.

Of the three targets of fraud, waste, and abuse programs, waste and abuse are the most difficult to identify and penalize. There are some inevitable scrap, waste, rejects, and nonproductive time elements in just about every human activity. To identify what part of waste is improper is a tough job. This is why positive incentives may yield much more improvement than negative ones. Identifying savings through waste reduction programs, cost reduction programs, and value engineering activities may be much more fruitful. Abuse can only be effectively reduced by hiring executives and employees with high integrity and honesty. Fraud is both detectible and punishable when laws are enacted for its prevention. But all three—waste, fraud, and abuse—are negative attributes that are as difficult to detect and penalize as are the disinterest, neglect, and apathy that allow them to exist in the first place. We prefer an approach that would eliminate, or at least reduce, the three culprits of fraud, waste, and abuse by replacing negative with positive incentives. Emphasis on penalizing industry for fraud, waste, and abuse should be matched or exceeded by practices that encourage frugality and honesty. One way to encourage these positive attributes is to give government managers the freedom to reward those who have exhibited these attributes.

Several years ago, the idea was advanced in government circles that industry should provide warranties in major systems and projects for the government in the same manner as for commercial products. A set of laws was passed and regulations were issued to require warranties on major systems. These provisions turned out to be both expensive and difficult to enforce. Mr. Ken Kastl of Boeing Military Airplane Company, commenting on the Weapon System Warranty Act at a 1985 government/industry cost estimating workshop,[19] said, "There is the possibility of increased cost to the government as a result of the prime (contractor) pyramiding the warranty down to his subs." He continued: "Increased costs are almost sure to arise because of efforts involved in warranty administration." The increased costs are passed on to the government in the form of a higher contract price, since the government is the only customer for many of the firms affected.

The Warranty Act states that, unless waived or exempted, the military departments and defense agencies may not enter into a production contract for a weapons system without obtaining three separate warranties: (1) the contractor must guarantee that the system conforms to the design and manufacturing requirements stated in the contract; (2) the contractor must guarantee that the system is free from all defects in materials and workmanship at the specified time of acceptance or delivery; and (3) the contractor must guarantee that the system will conform to the specified essential performance requirements. All of these guarantees are nice to have, but are we really getting better and more cost-effective systems and projects than we did without the warranties?

Mr. Kent R. Morrison of Crowell and Moring pointed out numerous pitfalls in the Warranty Act in a paper presented to the National Estimating Society in June 1985.[10] He said that shifting risks to contractors has questionable benefit as a matter of public policy. He cited the failure of the "total package procurement" contracts of the 1960s and the "total system responsibility" clauses of the 1970s regarding defects in the government's own design and specifications. Both of these earlier "reforms," it is noted, did not employ phased procurement of meticulously defined work but attempted to shift the responsibility of resolving unknowns in the entire job to the contractor.

Development costs increased under the Warranties Act because of increased engineering design and reliability analysis effort, more tests to demonstrate essential performance characteristics, and a longer weapons system acquisition cycle. The influences of the Warranty Act, such as a disincentive for concurrent development and production because of technical performance uncertainties, and disincentives to use high-risk technologies because of unknowns, were good in an overall sense because they helped avoid those common destructive program tendencies caused by competitive optimism. Government cost analysts found, however, that the government itself usually ended up paying the cost, whether the risk was taken by the contractor *or* the government. Problems such as contractors in a competitive position failing to put sufficient funds in proposals to adequately cover warranty costs; warranty costs being loaded into indirect costs partially allocated to other government contracts; prime contractors passing risk down to subcontractors at a substantially higher cost; the difficulty of proving cause of failure if the system is destroyed; and hidden litigation costs were all surfaced by cost analysts as being counterproductive to the original intent of the Act, which was to provide high quality systems at a reasonable cost.

Management incentives that encourage excellence in product quality and performance are likely to be more effective than more warranty laws and regulations. As anyone who owns a toaster, computer, or electric toothbrush knows, the warranty is only as good as the company that stands behind it. If the company has a desire and attitude to please the customer and provide an excellent product, it will automatically and persistently stand behind its product even to the extent of replacing it or returning the customer's money. Better than replacing faulty components after they become defective is supplying an item that won't break in the first place. Companies that do this should be on the preferred list in government contracting just as this type of company is given preference in the purchasing and procurement habits of private companies and individuals.

Auditing as a Constructive Process

The reforms, directives, laws, policies, procedures, and practices that have been enacted in past years—including those such as the False Claims Act Amendments of 1986 not previously mentioned (the original act was passed by Congress in 1863)—require costly, in-depth management, administration, oversight, implementation, and auditing to ensure that they are followed.

To audit its procurement activities, the government has an extensive hierarchy of auditors, inspectors general, and oversight functions. Each executive branch agency has its own internal auditors and investigators, both within the agency or service and at field installations. In addition to the internal auditors and investigators, departments have independent audit agencies such as the Defense Contract Audit Agency (DCAA) to oversee the entire department's procurement activities, to perform investigations, and to conduct detailed field audits at government installations and contractor plants. Groups of DCAA auditors are located in most large government contractor plants. The DCAA alone employs approximately 4,600 people in more than 430 offices throughout the continental United States and overseas. About half are in area offices in the larger metropolitan areas, and half work full-time at the plants of the larger defense suppliers. The agency operates its own school called the Defense Contract Audit Institute in Memphis, Tennessee, on the south campus of Memphis State University. The institute offers twenty-two courses in subjects such as cost allocation, statistical sampling, accounting, and audit principles and trains nearly 3,000 auditors in DCAA and other agencies annually.

The largest audit and investigations agency of all is the U.S. General Accounting Office (GAO), which is an arm of the Congress. The bulk of major systems audits and investigations are done by the National Security and International Affairs Division of GAO, but other divisions oversee construction, information systems, and civil projects such as mass transit, dams, highways, and public buildings. Since GAO is an instrument of Congress, it is not a line organization and does not have a direct controlling input to the heads of departments and agencies. In order to have its recommendations adopted, the GAO must convince the agencies to adopt them or persuade Congress to enact legislation to carry them out.

Hence, there are at least three layers of auditing within the government for major acquisitions: (1) internal audits, (2) external audits from the executive branch, and (3) GAO audits and investigations. These three layers do not include a fourth layer of audits which the industry itself requires to comply with various company rules and government statutes. Phased deregulation of the procurement process would include a reduction of audit layering. An ideal task for audit agencies would be to assist the planners, cost analysts, and cost estimators in developing efficient procedures and accurate cost estimates for future programs. Thomas Hays, Auditor of the State of California predicted that "in the future, there will be more reliance on audit data to help in decisionmaking." Auditors, in their oversight function, have developed valuable experience that could be used to estimate more realistic costs for future programs and increase efficiency. Let's put this experience to work for us.

We suggest also, as mentioned earlier, that auditors look for successful, beneficial, and productive actions by agencies and contractors as well as those that are wasteful, nonproductive, or illegal. In our view, auditing should become a constructive process for encouraging outstanding performance rather than just one to identify wrongdoing. An accent on the positive by auditors will help identify and reward good performance and will place more emphasis on identifying and promoting the practices, procedures, and policies that result in success. These practices can then be transferred to other not-so-successful programs to help spread excellence throughout government procurement.

A Vital Career Position: The Contracting Officer

The contracting officer for major systems acquisitions, who is usually a part of the project management team through direct assignment, co-location, or a matrix organization, is the person who must interpret the myriad procurement laws and keep the project manager and the agency within the law. Contracting officers, whose jobs are affected more than any other group by the proliferation of procurement regulations, come from diverse educational and professional backgrounds and depend heavily on lawyers for help.

Contracting officers must have a mixture of law, business, and technical backgrounds in order to understand the complexities of high-technology weapons system acquisition. According to a recent survey by the Defense Manpower Data Center,[13] the Department of Defense alone employs more than 24,000 contracting specialists, approximately 85 percent of whom are civilians. Statistics show that 54 percent of DOD's civilian contract specialists have college degrees; 28 percent have business degrees. Thirteen percent of the civilian contract specialists responding to a survey said they had neither a college education nor post-high school business courses, and another 33 percent have taken college level courses but identify their major at their highest level of study as something other than business. Military officers must have degrees to receive commissions. On the overall, it is apparent that the present procurement work force is ill-equipped to cope with even the present degree of regulation, much less additional layers of controls that could be enacted in future sessions of Congress without prudent restraint (see the next chapter, "The Role of Congress"). A combination of procurement law simplification and elevation of contracting officers' position in the project management team through intensive career enhancement will result in remarkably improved project team efficiency.

In recent months an initiative to develop an "Acquisition Corps" of trained career procurement personnel has been discussed. The proposal includes a Defense Acquisition University to expand the available training and educational opportunities for procurement personnel. This university, which would be available for training technical and business members of project teams, would most likely be combined with the Federal Acquisition Institute and would encompass courses provided by many of the government training facilities listed in Chapter 3, "The Project Manager and His Team." This training initiative would be a vital link to train and retrain procurement professionals. The FY'87 Air Force Acquisition Statement said, "The contracting career field, for example, already has a turnover rate of 13 percent above other career fields. The Air Force projects a 20 percent loss of contracting professionals due to retirements alone in the next five years." Simplification and deregulation of the procurement process, coupled with an upgrading of the procurement work force through improved training, could work toward a more stable environment in that work force.

Stability in the procurement work force is just as important as stability in project financing and in project manager assignments. The complexity of current laws clearly requires experienced professionals to efficiently and legally carry out the procurement of major systems as well as other products and services required by the government. The contracting officer must be one who is permitted to exercise good judgment, experience, fairness, and flexibility in acquiring major systems and projects in the best public interest.

The contracting officer is the person who would benefit most from simplification of the procurement process. In many major system project offices, the contracting officer is the principal legal advisor to the project manager. He or she is the person who officially signs off on the procurement and, therefore, has an enormous responsibility to see that all laws, rules, regulations, and policies are followed meticulously. Adding new laws, rules, and regulations, or changing them continually, merely adds to the complexity of the tasks of the contracting officer and restricts flexibility of the project team in doing its job. Reduction of the number of constraints and simplification of procurement would permit the contracting officer to work more closely with the project manager as an integral member of the project team in exercising control and in managing the acquisition. And improving long-term career opportunities for contracting officers will provide greater motivation, training, and professional advancement opportunities for this important and key member of the project team.

Needed: Improved Scheduling of Proposals and Contracts

Industries going after government contracts for major systems have found that they have to spend enormous amounts of time and money to prepare proposals, negotiate contracts, and submit the paperwork required under the procurement laws. The *proposal* has become the principal marketing tool for major systems and prevails as the most important document leading to acquisition of a government contract. Proposals to the government for large systems are known to have taken up to fifty volumes of detailed technical, schedule, performance, management, and cost data, with the volumes averaging 1-1/2 inches thick.[20] Costs of preparing proposals range from the millions to the tens of millions of dollars for large systems, and the preparation of the proposal and its supporting engineering and cost data often starts months and even years ahead of the expected contract date. Simplification of procurement procedures, reduction of procurement times, and a more predictable environment and time schedule for procurements will help reduce the enormous costs of proposal preparation—which costs are ultimately paid by the taxpayer.

Government evaluators usually set up source evaluation boards the size of a small company and keep them in session for up to a year to wade through the massive volumes of material requested from and submitted by the major contenders. While the evaluation is going on, the bidders are called in to provide verbal and written clarifications, and they must provide a "best and final offer" prior to hearing the selection decision. Selection decisions are often delayed because of various political, social, or economic reasons. Even after the selection is announced, it may take months before the actual contract is awarded and the cash starts flowing in. As one can plainly see, competition in this manner is not for the economically deprived or fainthearted. As mentioned earlier, companies that already have a significant amount of government work are allowed to charge bid and proposal costs and independent research and development costs to their overhead, which is borne by all of their government contracts. [The U.S. government may be the only customer that formally helps its supplier develop new technology and then propose it to the customer (the government) for its next job or set of jobs.] Once placed in the category of a government contractor, as long as the company can continue to win competitions, independent research and development and bid preparation costs are allowed under its government contracts. Those not already in the category of federal contractor must compete with their own money. The government's costs of source evaluation, as well as industry's cost in supplying evaluators with additional data, could be significantly reduced by shortening and simplifying the acquisition process for each project phase.

Increasingly, the government is performing independent cost analyses and estimates as a yardstick to measure the credibility of the bidders' prices. Adjustments are made to the bidders' prices to develop "most probable costs" on which the selection decision is based. Hence, the lowest bidder is no longer a certain winner. Selection of other than the lowest bidder, however, places a tremendous burden on the cost analysts who made the adjustments because a bid protest may be lodged by the company having the lowest bid if it is not selected. The government's adjustments and selection must then undergo scrutiny by the GAO and Congress before the selection decision can be either confirmed or overturned. Protesting competitors usually have no qualms about bringing their Congressional representative in on the decision to put pressure on the agency to reverse the decision in their favor. Greater reliance on independent government estimates as impartial yardsticks, and wider acceptance of these techniques by both the industry and the Congress, could speed up the process while retaining credibility in achieving good quality at reasonable costs.

During the proposal evaluation process, government boards engage in meticulous tracking of cost, performance, and schedule bids against the requirements. The bidder must have supplied solid substantiation and complete traceability of cost to schedule and performance *except in the last few days of the evaluation period*, when the contractor is permitted to submit the "best and final offer." Because of the likelihood of transfusion of business and technical proposal content during competitive negotiations, a really serious proposal is often not made until best and final offers are submitted. Consequently, initial proposal budgetary and planning figures may be based on a terribly inefficient program approach. At the time of the best and final offer, the bidder quotes the lowest price possible, but there is often an absence of systematic correlation and allocation of the new bid price and disbursement throughout the job to be performed. It is usually a *"bottom-line"* bid which the government must either accept or reject. Cost traceability is lost; and, if the contract is slated to be a cost-reimbursement type, or if changes to enhance the system's performance or schedule are readily accepted, costs immediately begin to increase. Costs then tend to increase in cost-reimbursement situations until the project manager's contingency is consumed and an appeal is made to higher headquarters or to Congress for more funds. Unused contingency funds sometimes remain—but they are rare. If the Congress is sympathetic to providing more funds, as it has been in the past for favored projects, an atmosphere is created for repetition of the same cycle within the same project or within other projects in the same agency. Simplification of the procurement process could include elimination of "best and final" offers and their detrimental effects on continued cost credibility throughout the program. Let each bidder place his best and final offer at the start of the selection process, suitably backed up by in-depth rationale, and merely make traceable adjustments to cost, schedule, or performance factors during negotiation of the contract.

A principal function of the procurement process should be to assure that there are no project or contract overruns. That is, the procurement process should be designed to help the project costs stay within the projected budget. Project and contract overruns are cousins, but not really the same. Charles J. Gaisor of the Defense Systems Management College[21] defines a contract overrun as "the narrow circumstances where actual contract cost for a product or service, under the then-current specification, is greater than the then-current target or estimated cost." "Importantly," Gaisor points out, "overruns are never fee bearing. Other modifications (both within and out of scope) to a contract are fee bearing." A project's budget could be exceeded without having contract overruns by experiencing add-ons. Therefore, many project offices that point to the fact that they have had no contract overruns cannot do the same when it comes to project overruns. If more contracts are given to the same contractor to complete the same work, the budget-shattering effects of both types of overruns are the same. To help alleviate this problem, source evaluation boards and source selection officials must be given the authority and responsibility to include past contractor performance in meeting cost, schedule, and performance goals as a major factor in contractor selection. Contractor survival and growth motivations must be used effectively in encouraging meticulous financial projections, tracking, and management in both the proposal phase and the performance phase of programs.

In untold numbers of procurements, the complete procurement cycle (proposal request, proposal preparation, proposal evaluation, best and final offers, source selection, and even negotiation) has been completed and then the procurement is delayed, sometimes for extended periods because the money is delayed or no longer available. In these cases, the procurement cycle has imposed large costs on both the industry and government participants, and both are left hanging in limbo until funds are restored or other political or procurement issues are resolved. In a simplified procurement process designed to be efficient, effective, and economical, both the industry and the government participants must be given the assurance that schedules will be met in soliciting and awarding contracts. Industry should be notified in advance when requests for proposal are to be issued—and these issue dates should be met. Industry should also be notified in advance when selection announcements are to be made, and selection dates and contract award dates should be met. (In a recent contract awarded for NASA's space station, funds were cut after contract award, resulting in a stretch-out and an enormous total program price increase.) In our view, one piece of enforceable legislation that would replace many other nonenforceable laws, and that would eliminate much costly spinning of wheels by both industry and government, would be to require that funds be available and committed for the total job and that all pathways to a completed procurement are cleared prior to the release of a proposal request.

Much needed systematic application of common-sense approaches like these to source solicitation, evaluation, and selection, along with the application of risk-reduction and risk-elimination techniques, will gradually change the proposal/contract process; and simplification of procurement laws, rules, regulations, and policies will result in sizeable savings to both industry and government in proposal preparation, source evaluation and selection, and contract administration.

The Future of Government Procurement

Assuming that significant reforms can be made in long-range planning, establishing financial stability, increasing the tenure of project managers and their teams, and in simplifying procurement laws and regulations, we predict a gradual movement from the current cost-reimbursement type of contract to a fixed-cost, phased-procurement approach that is based on meticulous definition, estimating, and cost management. This simply means that programs and their resulting contracts must live within their means, not just for individual contracts but for overall projects. Only when this philosophy is fully adopted will the trend be toward greater realism in estimating and bidding, coupled with shrewd management trade-offs to stay within cost boundaries. *Compulsory* policies, laws, directives, and "initiatives" in this direction must be avoided, however. Forced evolution of common-sense practices has inevitably resulted in over-complication and delay rather than simplification and expediency. As Donald A. Hicks said in 1986: "We must avoid the temptation to try to solve our problems with simplistic solutions . . . many of which only unnecessarily increase our complexities." A natural transition to a more definitive type of contract will evolve as programs are divided into phases; meticulously analyzed, defined, and estimated with the newest state-of-the-art techniques; and carried out in an environment that encourages trade-offs to stay within budgets.

During the early non-hardware phases of programs, cost ceilings can be placed on manpower, material, computer time, travel, and administrative costs. As prototypes come into play, fixed-cost contracts employing design-to-cost engineering techniques can begin to develop products with the correct balance between austerity and performance. Soon, our federal acquisition system will be able to demonstrate, through real programs, that it indeed is possible to exercise the same type of wisdom, judgment, and restraint that has been adopted in the commercial marketplace through natural (not legislated) competition. Fixed-cost phased-procurement can be accompanied by selective deregulation, which, in turn, will result in naturally induced rather than manufactured competition. A less regulated, motivated industry can work hand-in-hand with the government to form a team that has a much improved opportunity to overcome our real commercial and military adversaries from abroad.

Many countries (Japan is one) have benefited from a long-term atmosphere of cooperation and teamwork between government and industry in large long-term projects. In the 1990s, industry and government will be forced to join hands to compete with other nations in a world market. As an example of emerging industry/government cooperation in the United States, *Contract Management Magazine*[16] reported in October 1986 on a unique landmark "extracontractual" agreement between an industrial firm and a government agency to decrease government oversight and contractor overhead. The industry was TRW, and the agency was the Air Force. Perhaps this is the type of entrepreneurial cooperation that should be encouraged between contractors and their sponsoring agencies. It happened spontaneously without a DOD directive, Congressional legislation, or an Air Force initiative. In the same issue, *Contract Management* reported that more than thirty of the nation's top defense contractors pledged to "abide by specific principles of ethical conduct as a result of the recent recommendations of the Packard Commission." The companies involved represented nearly half of all defense procurement spending in fiscal year 1985. Since that time, other government contractors have voluntarily added their names to those who embrace a specific code of ethics and conduct. Industry itself is beginning to realize that an adversarial relationship with the government and a tarnished public image are counterproductive to the United States position in world markets and to the security and general welfare of our country.

With all of the foregoing said, it can be seen that the government procurement process is enormously complex, flawed in a number of areas, and confusing. It is long overdue for simplification and phased, selective deregulation. But what is the *desired* environment? How should the process work? And what should we do that differs from what we do today?

The desired procurement environment is one in which an atmosphere of mutual trust and respect exist between the industry producing the projects, systems, and services, and the government which must manage the whole process. We see the need for a vastly simplified set of controls, an atmosphere where excellence in achieving goals is rewarded, and where factors that encourage deception and circumvention do not exist. The environment must be one where an agency is not afraid to reward excellence by selecting other than the lowest bidder as long as that bidder is highly qualified to do the work at a reasonable price as judged by comparison with an independent government estimate; one where truly innovative ways of controlling costs are preferred to unrealistically low estimates; and one where the ability to deliver is considered more important than the ability to postulate beyond the achievable. In this new environment, legislation would only be used to impose *enforceable* criminal penalties for willful fraud. Ethics and morality policies would be generated by the industry itself and published in an ethics handbook, signed by every contractor.

In the new environment, as in any team effort, each industry should be able to see where it can fit into the dynamic national plan. Industry must be given the benefit of insight into national long-range planning—and must have sufficient visibility to be able to make reasonably sound projections of probable projects and workload well into the future. Industry, through its normal lobbying and Congressional representation activities, must have an input and voice into this very dynamic planning process, just as it does on a shorter term but sporadic basis today. Having had input into a soundly-based plan—one that is stable enough to make projections yet dynamic enough to adapt to new technologies—the corporations that support our major systems and projects can then make their own spending and capital acquisition plans with greater confidence in what the future holds.

The desired environment also includes increased teamwork rather than increased enforced competition. The term "free and open competition" has really been turned into forced competition—and we are rapidly finding out that this is inefficient and leads to poor quality. The new environment must enhance the ability of excellence to be rewarded rather than punished. Companies that do good work have every right to expect to be favored in future competitions, and those who cannot provide high quality and meet their commitments must not be permitted to continue to deliver inferior products, services, or major systems.

How, then, should this process work to produce an environment of excellence, teamwork, achievement, and quality in our major systems and projects?

First, we must recognize that major systems and projects fall into distinct phases and jobs, and that a major system's life cycle is comprised of these distinct phases and jobs rather than one continuous process. The acquisition of a major system can be subdivided into solicitations for ideas, solicitation for designs, solicitation for development and testing, and solicitation for production. Not all companies can do all of these jobs well. There are some companies that excel in the idea realm, and may have little capability for detailed design, development, or production. These are often smaller, tightly organized teams of systems analysts who have the ability to conceive, analyze, and formulate ideas and to envision how these ideas can be best used and put into practice. There are other companies and teams that can take ideas and create detailed plans, designs, and cost estimates. There are still others that excel in development, testing, and production. For each major system, the phases employ different people or different companies to do the part of the job they can do best. Even better, however, is a narrowing-down process from many competitors to fewer; then to one producer.

We have previously shown how and why the process must work in sequential steps to be most efficient. Now we can see that this stepwise process permits the early introduction of more players whose talents can be used where they are most capable. This process permits the introduction of more checks and balances in the stepwise systems acquisition process than in the early assignment of the whole job to one company based on a solicitation at the beginning with very little if any detailed definition of the job. At the same time, implementing the new environment must recognize that specificity can and must increase substantially in each phase. This is directly counter to past tendencies of opening up the definition for new ideas at every phase and helps prevent the infusion of unrealistically high performance goals before they have undergone their trial by the fire of analysis, simulation, design, and testing.

A case in point is the Strategic Defense Initiative or "Star Wars" set of programs, some of which are projected for operation up to twenty years into the future. To do the difficult, time-consuming, front-end creative work, a whole industry subset consisting of several hundred small, closely knit, highly competent, flexible companies has evolved. These companies do the conceptual thinking, research, analysis, simulation, and preliminary design work required to bring ideas from their germination and embryonic state into systems and projects that are expected to work when put into production, and that will be asked to perform in a long-term operational environment.

These small, efficient teams are really the architect-engineers of future defense and space programs, and they have counterparts for other high technology systems and projects in fields such as transportation, communications, and information technology. Many of these companies can do final as well as preliminary design, laboratory scale model and full-scale testing of concepts, and even some early prototyping. These functions can also easily and economically be performed by small, flexible teams within the industry giants who will be charged with development, testing, and production of the mature products. Small, flexible, highly competitive industry teams are beginning to look a lot like what David Packard envisaged as "centers of excellence" for the conception of high technology systems, and can do this job without committing huge sums of personnel, resources, equipment, and facilities. Professional service, research-oriented, non–hardware-producing, highly skilled think-tanks, armed with high-tech software and some of the best brains in the country, can be likened to teams of well-bred ponies that can be easily matched to specific definition tasks. Their small size, diverse locations, and low inertia make them highly adaptable to a rapidly changing and advancing technology. They can be teamed as required on short notice through small definition contracts without committing huge resources.

Once the ideas for major systems have been conceived, analyzed, compared with other concepts, narrowed down to the most promising, and tested through computer simulation and then in the laboratories, they are ready to move into the next phases of full scale development and production to meet stringent operational requirements. It is here where the giants of industry can fruitfully come into play with their mammoth production plants, huge test facilities, and efficient assembly lines. Tooling up this large industry for "Star Wars" volume production, for example, can be done on a selective basis as systems and projects become more fully defined. The large industries must be kept abreast of the embryonic and evolving concepts through their independent research and company sponsored programs in order to do the applied research and production planning needed to potentially equip their facilities for projected rates of development and production. But they need not commit large capital expenditures until they have been able to match production and manufacturing technology to the sophistication of the very products and systems that will be produced. The affordability of the Strategic Defense Initiative, as well as other advanced programs such as the aerospace plane, NASA's space station, and high-tech communications and transportation systems depends to a large degree on how well the mix of small, medium, and large companies can be used by implementing a simplified, selectively deregulated procurement process.

In the meantime, all industrial participants—including small, medium, and large companies—are finding an increasing need for effective cost analysis techniques to assure both themselves and the taxpayers that whatever course of action taken is the most cost-effective. A small firm we visited in 1980 that was just forming to perform high-tech work made the statement that "we don't need any cost estimating." Now, eight years later, they are scrambling to get their hands on all the information they can find on cost estimating, cost analysis, and cost data. With the advent of Gramm-Rudman and other budgetary restrictions, cost consciousness has now filtered down to very low levels in government and industry in a way that it never has before, and the new environment will include the philosophy of "do it within the budget."

In working toward the new environment, natural rather than induced competition will produce excellence just as it always has. It is an environment where you "won't get penalized if you only get one bidder." It will reward realism, cross-checked by independent government estimates, rather than overoptimism. It will recognize the cost-effectiveness of cutting off performance for this generation system at a point before the cost/performance ratio becomes exorbitant, but it will permit introduction of higher performance in the conceptual stages of the next generation system.

How does this environment differ from what we have today? First, we will be assured, in our selection process, that we have the authority, intent, and charter to select and reward the most dedicated, honest, capable, innovative, and highest quality company rather than merely the lowest bidder. No longer will we tend to believe the lowest cost projection or the most optimistic promises. Second, we will keep track records to make it easier to reward those who meet originally proposed performance, schedule, and cost goals. Third, we will openly include industrial base and technology base (geopolitical) factors in solicitations and selections. For example, those in labor surplus areas who are favored will know this prior to the solicitation. The considerations usually included as "other factors" in a source selection, those not previously advertised or graded, will be more visible.

In the past, many a contractor has had an excellent proposal—equal at least in technical and management quality to a competitor—only to be "selected out" because of "other" factors. Had they known the outcome to begin with, they would never have bid on the job, saving their valuable bid resources for more promising prospects. Fewer bidders will appear on the "competition scorecards" and greater selectivity will speed up the procurement process and will make it fairer and less costly to all. Fourth, the new environment must relieve the procurement process of the heavy burden of socioeconomic objectives. It is becoming far too expensive and time-consuming to administer social reforms as part of our acquisition process. The prices of already expensive systems and projects need not and should not bear this burden. Fifth, the reward should be more government business for those who can *live up* to their claims, not just to ones who can *make* the claims. Within the bounds of fairness, those who excel should reap the benefits. And sixth, the new environment will permit unique, highly specialized, and competent industrial teams to develop who can do specialized tasks.

It is not always necessary to have two or more bidding just to ensure an excellent product. A current process that now leads to "leveling" of competence and discouragement of excellence at the price of developing more players than can ever make the team should be replaced by one which will reward competence and permit those who excel to be first in line when it comes to new work.

These environments, implementation suggestions, and outcomes may seem to be extreme to many in the ranks of government procurement, but a system that has been given the opportunity to result in excellence for far too long has continued to fail to guarantee excellent results for all of our nation's projects and systems. Positive and effective measures are needed to restructure the process to parallel more closely the free market that has made this country great. Further regulation, restriction, and control of the procurement process will only continue to tie the hands of government managers so that they cannot fully utilize the creativity, drive, dedication, innovation, and persistence of American industry in providing major systems for its own government.

Endnotes

1. Jacques Gansler: *The Defense Industry,* MIT Press, Boston, 1980.
2. Aerospace Industries Association, *Aerospace,* Aerospace Industries Association, Washington, DC, Spring 1984.
3. Institute of Cost Analysis, *Journal of Cost Analysis,* Institute of Cost Analysis, Washington, DC, Spring 1985, Vol 2.

4. Packard Commission: *A Quest for Excellence: Final Report of Packard Commission, 2 vols.*, Packard Commission, Washington, DC, June 1986.
5. Merton J. Peck and Frederick M. Scherer: *The Weapons Acquisition Process: An Economic Analysis,* Harvard University, Boston, MA, 1962.
6. John J. Kennedy: *Incentive Contracts and Cost Growth,* Air Force Business Research Management Center, Wright-Patterson AFB, OH, October 1983.
7. John J. Kennedy: *The Effectiveness of the Award Fee Concept,* Naval Office for Acquisition Research, Washington, DC, June 1986.
8. Office of Management and Budget: *Executive Order 12352 on Federal Procurement Reforms: A Reform '88 Program,* Office of Management and Budget, Washington, DC, February 1984, M-84-7.
9. Military Business Publishers, *Military Business Review,* Military Business Publishers, Virginia Beach, VA, July-August 1986, Vol. 2, No. 4.
10. National Estimating Society: *National Estimating Society Conference '85,* National Estimating Society, Washington, DC, June 19–21, 1985.
11. Commission on Government Procurement: *Major Systems Acquisition—Final Report,* Commission on Government Procurement, Washington, DC, January 1972.
12. Office of Federal Procurement Policy: *Competition in Contracting Act of 1984 Analysis,* Office of Federal Procurement Policy, Washington, DC, October 1984.
13. National Contract Management Association, *Contract Management,* National Contract Management Association, McLean, VA, December 1986, Vol. 26, Issue 12.
14. General Accounting Office: *Legislative Recommendations of Commission on Government Procurement 5 Years Later,* General Accounting Office, Washington, DC, July 1978, PSAD-78-100.
15. U. S. Congress, House: *Omnibus Procurement Reform Amendment to HR 4428, DOD Authorization,* U. S. Congress (House), Washington, DC, August 1986.
16. National Contract Management Association, *Contract Management,* National Contract Management Association, McLean, VA October 1986, Vol. 26, Issue 10.
17. Department of the Air Force: *Air Force Acquisition Statement FY87,* Department of the Air Force, Arlington, VA, 1986.
18. Joint Financial Management Improvement Program: *Proceedings of the Fifteenth Annual Financial Management Conference 1986,* Joint Financial Management Improvement Program, Washington, DC, March 1986.
19. L. S. Ritchey: *ASD/Industry Cost Estimating Workshop, October 1985,* United States Air Force, Dayton, OH, December 1985.
20. Rodney D. Stewart and Ann L. Stewart: *Proposal Preparation,* Wiley, New York, 1984.
21. Defense Systems Management College, *Program Manager,* Defense Systems Management College, Fort Belvoir, VA, July-August 1983, Vol. XIII, No. 4.

CHAPTER SEVEN

The Role of Congress

In a democratic society there is a trade-off between the special interests of each citizen and the corporate interests of the citizens as a whole. This fundamental trade-off manifests itself in the body that makes the laws that set the policies of government, the U.S. Congress. The major systems, projects, and programs instituted by our government have become visible entities where the brunt of this conflict takes place, and their economic, social, and industrial importance have a profound effect on the electoral process. The Congress has tended, in the past, to focus more on each major system or project in the light of its impact on its own constituency than on the total mix of major systems and projects needed to accomplish national goals. The focus tends to center on whatever system or project is under consideration at the moment rather than on the available resources within the total mix of major systems. The actions we are advocating are those designed to create and maintain a better balance between the two types of interests, not to have the corporate interest overwhelm the private interest as is the case in communist countries. The major problem is that the Congress has tended to look at the short-term effects of major systems, and to view major systems as isolated entities rather than as an optimum mixture of projects that fit well together in an overall picture to meet long-term requirements.

Although representatives and senators have individual terms of two and six years respectively, a study cited earlier shows that the average years of service of the average member of Congress equals or well exceeds the life cycle of most major systems. The objective, then, is to encourage the Congress, as a body, to adopt a longer range perspective in regard to major systems and projects—one that transcends the individual terms of individual members of Congress by factors of three to ten. How can this be done while still recognizing and honoring the need for each representative or senator to reflect the interests of his or her electorate? This can be done, first, by providing the Congressional committees and subcommittees with credible, in-depth, long-range planning that shows how specific systems and projects fit into an overall plan to meet national goals. It can be done, secondly, by increasing both Congressional and public awareness and knowledge of the benefits of continuity and stability on the major systems acquisition process. These goals require each citizen and each elected official to balance, but not overwhelm, strong special interests with strong patriotic interests that will help achieve efficiency, economy, and effectiveness in all of our major systems and projects. It can be done, thirdly, by permitting the Congress to authorize, appropriate, and commit funds that fit within the budget for multiyear phases of programs. Representatives and senators should be confronted with and participate in the evolution of a dynamic long-range national plan that is compatible with projected budgets, and to weigh this plan against special interest pressures. With the current and continuously visible bigger picture available, individual legislators are less likely to advocate a specific system or project without considering the impact on the overall long-range plan or available resources.

One of the reasons for the complexity and detail in the federal budget is that agencies and the OMB are required to subdivide costs of major programs into expenditures in each state and then into each Congressional district. This requirement was imposed by Congress to permit individual senators and representatives to identify how many workers are affected and how much is to be spent in their state or Congressional district. This is a worthy effort, but soon it becomes a contest between the legislators to be sure they get "their share" of the federal budget channeled into their home areas in which reside their constituents. There will be little movement toward attempting to save huge sums of money in our major projects by improving program stability until the Congress and the citizens that they represent become more aware of the devastating effects of unbridled parochialism. Pressure by citizens, businessmen, employees, unions, companies, and chambers of commerce to bring work (dollars) into their Congressional areas becomes intense during the early budgeting phases of major systems. This geographical "allocation" of funds presumes that certain companies will get certain contracts. Source evaluation boards usually recommend the most efficient contractor, but pressure is subsequently put on source selection officials, sometimes subtly and sometimes not so subtly, to select contractors in specific states or districts. Hence, throughout this process there is much attention to which contractor in which state or district gets the contract with little overall pressure to award the contract or subcontract to the company that will provide the best product at the lowest cost.

Pressures to award contracts to companies in specific locations become particularly intense when the area's unemployment is high or where failure to award the contract will result in loss of jobs or adverse economic impact. Thus, the process of systems procurement frequently takes on the appearance of spreading the wealth rather than selecting the most efficient contractor or subcontractor. In a democracy, it is not likely that these pressures will be eliminated or even substantially reduced, but macro-planning could *include* these potential influences to reduce the impact of last-minute parochial fights, avoid obvious geopolitical conflicts, and permit selection officials to make courageous decisions in favor of cost effectiveness. To generate an atmosphere where this is possible, we, as citizens, must become aware of the detrimental effects of regional pressures to award contracts, and we must convey these concerns to our legislators strongly enough to force more serious consideration of national goals in the legislative process.

A definite stabilizing factor in the Congress is the use of a hierarchy of committees and subcommittees consisting of membership from both major parties and supported by permanent staffs. The Senate has twenty-nine committees and ninety subcommittees, while the House of Representatives has thirty-six committees and 156 subcommittees. At last count, the Senate had 796 full-time committee and subcommittee staff members, while the House had 1,310 committee and subcommittee staff members. About 25 percent of the committee and staff activities in both houses of Congress are directly related to systems acquisition. Committees have a substantially large representation of representatives or senators: the average is about 15.7 members in the Senate and 31 in the House. Subcommittee membership numbers average 7.6 and 12.5 respectively. Senators participate, on the average, in eleven committees and subcommittees, while Representatives average seven. From these averages, it can be seen that, even though Congressional terms are shorter than most major projects, stability in major systems policymaking and implementation in Congress could be expected to be good because of the overlapping membership in committees and subcommittees as well as the continuity brought about by the employment of permanent full-time staffs.

Relationship of Congress to the Project Manager

Despite the potential for high stability and continuity in the Congressional guidance of major systems and projects, the typical project manager has a very different view of Congress. The project manager has the distinct impression that he or she has a "silent" but giant partner in carrying out his or her project. This silent partner can authorize additional funds; reduce, delay, or terminate the project; change system specifications; affect project schedules; influence contractor selection; or terminate or delay the project at any time—all with very little prior notice. If this silent partner supports the project throughout the project's lifetime, the project manager will have clear sailing ahead. Well planned, conceived, and estimated programs that have had the long-term support of this silent partner have generally succeeded. Those that have not had the support of Congress have usually failed to meet original cost, performance, or schedule goals.

Throughout the years, a very special relationship has developed between the project managers and the members of Congress. The principal reason for this relationship, in our view, is that the Congress must hold someone in the executive branch responsible for every program. Since he or she is the senior official on the project, the project manager is often called to testify before committees and subcommittees, asked for written depositions, and requested for information regarding the project. It is for this reason that we treat the Congressional side of the Congress/project manager relationship in more detail than the Congressional interfaces with the President, the Cabinet members, and the agency administrators or the service secretaries. If the Congress is where the buck starts, the project manager is where the buck stops—as far as making and sticking to major program commitments is concerned. Even though the project manager may be far removed organizationally from the members of Congress (in a different branch of the government and often insulated by numerous supervisory layers within the executive branch), he or she is most often singled out as responsible to the Congress and to the public for successful completion of a project. We have discussed the duties, responsibilities, desirable characteristics, and tenure of project managers in detail in Chapter 5. Now, before exploring in more detail the role of the Congress in major acquisitions, let us search through the historical data on major systems to locate succinct descriptions of this very special and important relationship.

It probably can and most likely will be said that one can glean almost any statement from the massive archives of information on government procurement to prove a point. But when one goes back to a very specialized study on the government's treatment of major systems, a study in which millions of dollars and thousands of hours of time of dedicated men and women were spent, one would expect credible findings, conclusions, and recommendations. This relationship between the project managers and the Congress was studied thoroughly by the same group that has been referenced before in this book. We return to the report of the Major Systems Acquisition Study Group of the Commission on Government Procurement. A special task team of the major systems group dealing with project management, headed by Orville Enders of General Electric, developed, in 1970–1971, the recommendations shown in Appendixes B: "The Organization and the Role of the Project Manager," and C: "The Role of Congress in the Management of Major Systems Acquisition."

For brevity, only the findings, conclusions, and recommendations of the study of these two subjects are presented. An interested researcher will also find enlightening the accompanying discussion included in the study group's final report[1] in the library of the Federal Acquisition Institute in

Washington. The two most revealing synoptical statements from these two exhibits that bear on the role of Congress are as follows:

> From Appendix B: *Although the role of the project manager has gained wide acceptance and recognition, it is quite evident that many of the management responsibilities upon which program success depends is guided and influenced at many levels of authority above the project manager. Such guidance and influence* starts with the Congress *and extends down through the executive with many layers of staff and management involved [emphasis added].*
>
> From Appendix C: The Congress is by far the most significant potential force to improve the management of major systems acquisition *[emphasis added].*

These two study results, chartered by the Congress itself, place the responsibility for improvement of the major systems acquisition process squarely on the shoulders of Congress, and conversely, by implication, trace the responsibility for not setting or meeting goals and objectives back to those same shoulders.

Is it true that Congress must be held ultimately responsible to the American public for timeliness, efficiency, economy, and effectiveness in our major systems and projects? If not, whom do we hold responsible? Why is the President not the one who has final say-so on government projects? The answers to these questions lie simply in identifying the source of money to fund our major projects. The body that controls the purse strings is the one that exercises the final control. Congress has control of the purse strings and, in the past, has not hesitated to use this control. Congress is where the money for the project is approved, and in this control the Congress assumes the role of the *parent organization*. The citizens of our country are the users or "clients" that look to the parent organization to see that the needs are met efficiently, economically and effectively.

Responsibilities of the Parent Organization

Responsibilities of the parent organization (Congress), the clients (end users or beneficiaries), and the project manager can be seen clearly in the results of what is believed to be the largest and most comprehensive project effectiveness study to date. This study, performed by Bruce N. Baker, David C. Murphy, and Dalmar Fisher, entitled "The Impact of External Factors and Conditions upon Project Success," *1974 Proceedings of the Project Management Institute*, San Francisco, October 19–22, 1975, as reported by Cleland and Kerzner[2] found twenty-two project management characteristics associated with success.

A third of these project management characteristics had to do with the activities, policies, and responses of the *parent organization*. Baker and company listed characteristics such as: commitment to established schedules;

enthusiasm; commitment to an established budget; commitment to technical performance goals; and lack of legal encumbrances from the parent organization as keys to project success. The remaining determinants of project success included responsibilities of the project manager and of the client or end user. Frequent feedback from both the client and the parent organization were recognized as needed in successful projects, as was the availability of backup strategies and adequate control procedures, particularly those for dealing with changes. But it is the parent organization that must set the overall policy, generate and maintain enthusiasm for the project, and be willing to stand behind the project through thick and thin. But, even with its elaborate committee structure and staffs, does the Congress have the inherent stability required to see long-term projects through to completion?

As discussed in Chapter 3, the duration of major Federal programs usually exceeds the two-year and six-year terms of representatives and senators, respectively. Interestingly, though, the average time for representatives and senators in the Congress exceeds the total duration of most major programs because many are reelected for subsequent terms. An informed university political scientist told us that the reelection rate in the House is about 90 percent, and the Senate reelection rate approaches 75 percent. Using these turnover rates as an example, 56 percent of the senators and 53 percent of the House members would still be in office throughout the total duration of a twelve-year major system or project. This represents a majority over the twelve-year period. If a program or project were unanimously approved at its start, it would have *at least* majority support at its conclusion. In fact, with the limitation of two four-year terms for President, the Congress can exercise control over a longer span of time (twelve years) than can a single President (eight years).

With the above in mind, the Congress is uniquely suited to be the parent organization to exercise the "successful project" attributes of enthusiasm for the project or system: commitment to established schedules, established budgets, and technical performance goals.

Further, the Congress is more fully equipped than any other government body to comply with one of the key parent organization attributes of successful projects: the lack of legal encumbrances. Since the Congress originates the legislative controls over the procurement process, it has the full power to see that this important "successful project" criterion is met. Less regulation of the process would not only increase free and open competition as it has in other industries (telecommunications, railroads, and airlines, for example) but would remove encumbrances to project success.

In Baker's list of successful project attributes, the "client" or user of the system or project also has a definite responsibility for contributing to project success through client commitment to established schedules, established budget, and technical performance goals. The user, then, must insist that the parent organization carry out its responsibility to acquire the system in an effective and successful manner. (Not only is the problem us, as Pogo states, but the *answer* is us!). In case this doesn't cover you or me because we are not "users" of the system being promoted, developed, or produced, Baker has a final clincher in his listing of criteria for success: *enthusiastic public support*.

Returning to an interview with Jim Gregory, retired deputy contracts chief of the Army's Safeguard Systems Command,[3] partially quoted in Chapter 1, we continue with the following:

RDS: *Jim, you hear a lot of things about the cost of military systems being high. In your experience is the cost higher than commercial or is it equal to commercial? If it is higher, is there a reason for it? What is your experience there?*

GREGORY: *Well, if you're talking about something comparable, say like some of the municipal buildings that have been built, if you compare that with missile systems, missile systems costs certainly haven't been as great as have those municipal buildings. But if you're talking about the aircraft industry or automotive industry, yes. But of course we don't produce as many as they do. Missile systems are usually low density items. A few of them are very high density but most of them are low density. We buy annually. The quantities are not fixed. They vary from year to year because of the Congress. You never know what the appropriation is going to be from one year to another. Therefore there is a wide variance in programs. And that's going to increase cost. And, of course you can attribute it to a lot of things. If you look at the SAR [Selected Acquisition Reports], there are a number of reasons that they exceed the estimates: anything from bad estimating to runaway inflation. Of course in the 1970s inflation was a big problem. But not as big probably as a lot of people would like to say it was.*

RDS: *Some people have tended to blame the cost increases on inflation?*

GREGORY: *Yes, they like to blame it on that! But we had escalation clauses that covered a lot of abnormal escalation; and then, of course, when you negotiate the contract you would include escalation to the extent you could reasonably estimate. But still it was runaway inflation. But one of the problems in the missile business, I guess, is that the design continues to change because you have an evolving threat; and as the threat changes you want to be able to combat that threat. So you make major changes.*

RDS: *Even during production?*

GREGORY: *Oh, yes.*

ALS: *Does the Congress have any affect on that—I mean if they provide less money, then you have to change the design—or have you already been awarded that money?*

GREGORY: *The Congress doesn't have that much effect on change I don't think . . . technologically. Their changes are in programs—how much they're going to appropriate. And of course it is very sensitive to what's happening in the world at the time. There's not unanimity in Congress now on any defense program. Every program is under attack by some faction; and, therefore, you're not sure from year to year what's going to happen.*

RDS: *Does that itself increase costs?*

GREGORY: *Well, yes! As an example, on some of our major programs we have gotten as many as eight different proposals in one fiscal year. And this was because we didn't know what quantity we were going to buy until Congress finally appropriated funds and that can be very, very late in the year even though you would proceed based on a continuing resolution. We've had cases in recent years that the appropriation did not support the continuing resolution so we've had to terminate programs because of reduction in funds. In major systems, the politics and the world situation have a great affect whether we're going to buy butter or bullets. That's what the Congress listens to, what the pressures are at the time. In the early Reagan years, of course, Congress was giving him [the President] just about what he asked for. So . . . he did pretty well except the shopping list was greater than we could afford so there was still a problem: what can be bought with less appropriations? There wasn't enough appropriated for everything that the defense establishment would like to have. You can't have, at the same time, a 600-ship Navy; an I don't know how many wings Air Force—twenty or thirty, whatever it was; and a sixteen-division Army. Our system just won't support that. So you're going to have to pare it somewhere. The Navy has been more successful than the Air Force in that department. I guess there was more commitment to the 600-ship Navy.*

RDS: *Do you think that there are any laws that could be enacted or any kind of procurement regulation that could be changed or any planning implementation started or anything that could increase the efficiency or economy or effectiveness of major systems?*

GREGORY: *Of course they [the Agencies] continue to suggest to the Congress that they appropriate funds for more than one year. That would be a big help because then the contractors and the government would know exactly what they're going to buy for at least a short period of time, rather than changing every year. And, as I said, for one system we had eight major proposals in one fiscal year.*

ALS: *I thought we had a five year plan of some sort?*

GREGORY: *Defense does have a five year plan and we work toward that but we never accomplish that.*

ALS: *That's just dreams?*

GREGORY: *Yes. But it's a framework and it's the basis for defense plans.*

ALS: *But there's really nothing to reinforce it, to back it up with?*

GREGORY: *Not if the Congress won't appropriate the money!*

Not only does Congress delay funding or allow it to fluctuate, but it delves into detailed scheduling, quantities to be procured, personnel assignments, and systems specifications. As an example, the following is a brief excerpt of an article about the $20 million Continental Army Management Information System (CAMIS) that appeared in the November 21, 1986, issue of *Government Computer News:*

> *A fed-up Congress, angered by Army delays in building a critical system for management reserve forces,* took matters into its own hands, laying out specific guidelines and deadlines for the project.
>
> *In legislation passed last month as part of the Defense Department appropriations bill, Congress asked for a* detailed plan for the project to be submitted by March. *The directive also laid out several parameters for the Army to follow in designing the system [emphasis added].*

An article about the same program in a later issue reported the resignation of project manager Colonel Charles S. Miyashiro because his project remained unfunded and because *"recent legislation instructed the Army to put an officer at the level of General in charge of the project."*

Congress endorsed an intercontinental ballistic missile modernization program on May 26, 1983, as part of the Department of Defense Authorization Act of 1984 (Public Law 98-94). In this law, the Congress set, among other criteria, a weight limit of "33,000 pounds for the missile, including its guidance and control subsystems,"[4] and established specific deployment quantities and schedule.

These are examples of how the Congress can get deeply involved in personnel and technical as well as budgeting and resource decisions.

In the East Coast National Educational Conference of the National Contract Management Association (NCMA) in November 1986, Secretary of Defense Caspar Weinberger criticized the number of detailed, time-consuming reports Congress requires of DOD and "Congressional neglect of its broad oversight responsibilities in favor of extraordinary attention to minor matters."

Other specific examples of detailed Congressional legislation affecting programs appeared in a May 1986 GAO report,[5] *Strengthening the Capabilities of Key Personnel in Systems Acquisition*:

1. On an Army/Air Force joint tactical missile program, the joint program was dissolved in favor of dual (individual service) programs. During the joint phase *"legislation directed* one of two airframes used in pre-program technology demonstration."
2. In a Navy "inner zone helicopter" program, the Naval operational requirement called for one design solution, but the Office of the Secretary of Defense directed a competitive acquisition. After competition was underway, *Congress "directed a sole-source solution."*

Gary Hart, commenting on micromanagement by Congress in his book, *America Can Win,*[6] said: "But together, the [FY1985] authorization and appropriation bills changed 3,163 line items. Each change causes major perturbations in the Defense Department as contracts must be altered, programs slowed down or speeded up slightly, plans changed, and so on—all of which generate more meetings, paperwork, and committee compromises." And this is just for the Department of Defense budgeting bills. Imagine the magnitude of the changes to budgets to other departments and agencies as well and their ripple effect down to the project managers and contracting officers. Hart also said: "Many committee members like micromanagement because it frees them from dealing with major issues. They shy away from taking on major issues because if they do, they may get in trouble politically if events prove them wrong." Perhaps an overstatement, but if the Congress is going to become so deeply involved in the details, we, the citizens, should also hold them responsible for the major issues that affect the overall success of our major systems.

In early 1987, after the enactment of the Competition in Contracting Act and many other procurement reforms, and after the appointment of procurement advocates and their staffs in the Department of Defense and the services, *Contract Management* magazine reported statement by Jaques Gansler, president of the Analytic Sciences Corporation, as follows: [Quoted by permission]

> *The relevant question is where do we go from here now that we have acknowledged there are problems with the acquisition process. The system can be fundamentally reformed by structural changes aimed at enhanced program stability, continuous competition, a more integrated approach between the services, and increased contractor self-governance. Or, the system can be increasingly regulated and micromanaged by Congress. The problem today is that Congress has passed a series of laws that attempt both alternatives at once even though there is conflict between them.*

Needed: Pre-legislation Analysis and Simulation

The need is becoming increasingly evident for greater in-depth study, analysis, simulation, and testing of procurement laws before they are enacted. A Senate Armed Services Committee staffer told us that he is alarmed at the way bills are introduced and debated on the floor of Congress without proper analysis by the committees and subcommittees to determine the real long-term effects of the legislation. In the hurry to get things done, there has been a general breakdown of the system that precludes careful forethought and analysis prior to enacting procurement reforms into law. One problem is that there are so many procurement laws that it is very easy to enact one that is at cross-purposes with others. Instead of being analyzed, studied, and simulated prior to its enactment, the law is rushed into a bill with the idea that problems can be ironed out later in other laws. Post-legislative ironing out of problems is enormously expensive when compared to the costs of pre-legislative simulation, careful study, and systematic analysis of the law's long-lasting effects.

What is needed in the federal procurement lawmaking process is a phased approach to legislation very similar to that which has been proposed by Congress to apply to the development and production of major systems and projects: adequate analysis, simulation, test, and evaluation before "deployment." An after-the-fact analysis of the problems brought about by the law, such as those, discussed later in this chapter, found in the Competition in Contracting Act, is like shutting the barn door after the horse is stolen. It is often too late to correct irreparable damage and delay to important ongoing programs.

Scientists and engineering researchers perform crucial experiments in nondestructive or non–life-threatening situations with prototypes or disposable replicas of the object of research. Likewise, with a procurement process that involves the expenditure of many billions of dollars each year, only those practices and procedures that have been thoroughly analyzed, tested, and proven successful in nondestructive situations should be used on the live patient. Potential legislation should be tested for synergy, homogeneity, and consistency with existing laws before being placed into the body of procurement law. Such a process should involve analysis, legal studies, surveys, opinion polls, role-playing simulations, and projections of long-term impacts. A thorough cross-checking process with other laws, rules, incentives, and initiatives should be carried out to ensure that the new law does not work at cross-purposes with other laws. Thorough and searching investigations should be made into the *detailed* impact of the laws on all citizens, bodies, and groups, not just on the special interests that are advocating the legislation. This is the way the Congressional committee system is supposed to work. Although it is often outflanked, at its best it can perform this function. To carry this concept even further, real-time simulations with various affected groups could employ role-playing, group discussions, and even computer search or simulation to determine legislative, socioeconomic, and geopolitical interactions of laws *before* they are enacted.

The Congress needs to ask itself these questions before enacting more procurement legislation that affects major systems acquisition:

- Would the new law generate a truly beneficial long-term change in management philosophy or approach rather than merely add another organizational entity, requirement, regulation, or set of rules?
- Has an in-depth analysis been conducted to ensure that: (1) there are no conflicts with other laws, rules, regulations or initiatives; (2) the law is truly workable and beneficial; and (3) the law benefits *all participants* in the procurement process in producing more efficient, economical, and effective systems and projects?
- Is there a general consensus among those who have studied the federal procurement process in depth that the law will endure in the long

term under the wide variety of procurement situations in major systems acquisition?

- Does the law take into account the phased acquisition approach and the long-term nature of major systems acquisitions?
- Does the law have the potential of providing an overall cost savings to the U.S. taxpayer when both industry and government costs are considered?
- Has the law been tested in a variety of sample procurement situations and scenarios to assure its flexibility, reasonableness, and enforceability?
- Does analysis show that the law will speed up and simplify the procurement process?
- Is the law consistent with overall long-range national goals, objectives, and dynamic plans?

The report of the Major Systems Acquisition Study Group of the Commission on Government Procurement[7] observed that agencies want flexibility; yet they go to Congress for approval of programs, projects, and systems on a piecemeal basis. In other words, the Congress is being asked to do the master planning. Since Congress has no master planning arm except the executive branch,[a] it does only what it can do after the fact, handle each program on a case-by-case basis.

In Appendix B, the project management task team of the Major Systems Acquisition Study Group reported:

> *Congressional needs for project management information should be fulfilled by the appropriate agency of the executive branch instead of developing a separate technical and systems analysis capability within the legislative branch.*

The response of the Congress was that too often the project management information coming from the executive branch is only intended to bolster advocacy of a program. An independent perspective is necessary. However, the only agency that has sufficient information to fully evaluate, simulate, and test the effects of procurement legislation is the executive branch. There is one possible exception within the legislative branch—the General Accounting Office. In view of the valuable experience gained by the GAO after many years of investigating, auditing, and studying the acquisition of major systems, it would seem that this experience could be used along with that of the executive branch agencies to assist Congress in predicting long-term outcomes of its proposed legislative reforms. Indeed, the GAO has made many recommendations to the Congress that have

[a] The Congressional Budget Office is "required to develop a 5-year cost estimate for carrying out any public bill or resolution reported by Congressional committees" but has no detailed independent estimating capability and must rely principally on the executive branch agencies and the GAO for its inputs.

eventually found their way into law—some with and some without executive branch blessings. A joint effort, such as that now in progress by the Joint Financial Management Improvement Group, which uses both executive and legislative branch inputs, would seem both feasible and productive to develop durable, error-free, long-term procurement statutes. Generation of such statutes that would be acceptable to both branches of government would alleviate the necessity to enact new procurement legislation every year to correct the long-standing problems in federal acquisition as well as the problems generated by the previous year's legislation.

The objective of a pre-legislation analysis, simulation, and/or testing program would be to more fully inform individual senators and representatives about what they are voting on and to assure them that the laws have been fully researched and debugged prior to reaching a vote. The following excerpt from the November 1986 issue of *Contract Management* confirms what we occasionally suspected was true—but now someone has said it:

> *As the 99th Congress neared adjournment it accepted a House/Senate conferees' compromise version of a National Defense Authorization Act for FY 1987 that contains a number of procurement and procurement-related provisions. While most apply only to the DOD, some also apply to NASA and other civilian agencies. The House and Senate conferees spent nearly three weeks ironing out their differences about the act. When the conference version came to the House floor,* one member indicated the House did not have either the bill or the conference report: "We have only a press release put out by the committee staff describing what is in the authorization bill." *Most of the floor debate in the House centered on the rules governing consideration of the bill and whether funding should be provided for the T-46 trainer aircraft. In the Senate the major controversy centered on the T-46 [emphasis added].*

We made a special call to one of the committee staff members to determine if it could actually be true that our representatives debate or vote on a law without even reading it. He verified that this could very well be true. This particular bill mentioned above contained some twenty important procurement initiatives, such as testing and evaluation of major weapons systems and munitions programs, increasing the use of multiyear contracts, and limitations on the use of undefinitized contracts.[b] It is not that individual members of Congress need more detailed information, but they need it in a thoroughly developed, credibly summarized format and on a more timely basis. The committee staffs hasten to say that members of Congress are not deprived of information, but that they are involved in such a large volume of legislation that they may not be able to digest all that is needed in order to become fully informed on a specific procurement issue.[8–10]

[b]An undefinitized contract is one under which work proceeds without a schedule and/or cost formally agreed upon. Letter contracts and undefinitized changes are included in this category.

An Example: The Competition in Contracting Act

As a case in point, Colleen A. Preston, counsel for procurement policy for the House Armed Services Committee, in an article in the summer 1986 *National Contract Management Journal*,[11] pointed out that despite the number of beneficial changes brought about by the Competition in Contracting Act (Public Law 98-369) and in related procurement legislation, adoption of this detailed and comprehensive law resulted in considerable confusion. In her article, "Congress and the Acquisition Process: Some Recommendations for Improvement," she cites the following impacts, which, while not necessarily created by adoption of the Competition in Contracting Act itself, were encountered during efforts to implement the law:

- The recent proliferation of procurement laws such as the Competition in Contracting Act, coupled with the fear of reprimand or ridicule for making mistakes, has tended to "remove any semblance of decision making" in the bureaucracy and to cause procurement officials to "strictly adhere to rigid regulatory guidance regardless of the wisdom of that action."
- The phase-in plan for the new law was inadequate. Officials were confused as to *when* to implement various provisions of the law.
- Implementation of the act resulted in increased procurement lead time. For example, the agencies added extra reviews, not necessarily required by the law. (One Air Force activity reported average lead-time increases from 183 days to 219 days for competitive awards and from 108 days to 181 days for noncompetitive awards.)
- The law made it more difficult and time-consuming to award a sole-source procurement even when that action was in the best national interest.
- Delays were caused by a requirement to advertise procurements in the *Commerce Business Daily*.
- Longer evaluation times were necessary because more bids had to be reviewed.
- The pressure to find alternative sources seriously affected product quality. Some suppliers provided counterfeit, foreign-made, or defective parts. Others were merely a representative or outlet for many manufacturers. Still others bid only on contracts that had a first article test requirement and then defaulted on the contract after receiving progress payments.
- Interpreting the requirements of the act has been difficult.
- The act does not clearly provide for excluding sources that do not comply with the "Buy-American" Act.
- The act does not authorize excluding a company based on an organizational conflict of interest.

- The act penalizes procurement from blind, handicapped, or prison industries by classifying these procurements as noncompetitive.
- The act does not clearly encourage awards based on life cycle cost.
- Interpretation of the act by agencies has resulted in the award of contracts based on slight savings in labor costs without adequate attention to the technical merits of the proposal.
- Bid protests (submitted to the General Accounting Office and the General Services Administration) and litigation are expected to increase because of the act's provisions.
- There are no time limits specified for protesting contract awards.
- One company can stop the entire procurement process by merely mailing in a bid protest. The protest must be resolved before the procurement can continue.
- The government bears the expense of protest procedures.
- Protests can be upheld due to the government's violation of a statute or regulation even though a protestor has not sustained any injury. (Any violation of a law or statute, no matter how inconsequential, could become the basis for invalidating a procurement.)
- The act has resulted in an extraordinary increase in the number of bid sets and technical data microfiche cards being requested by potential bidders with no corresponding increase in bidders. Information brokers are "building libraries of technical data at the government's expense" and selling information to clients.
- Poor descriptions in the *Commerce Business Daily* by agencies implementing the act are causing more companies to request data than actually expect to bid.
- Breakout of procurements results in additional costs when the government must act as a prime contractor rather than purchase through one prime responsible for all aspects of the system.

Although many of these discontinuities in the law affect the purchase of small parts and services, major systems and projects that use these parts and services are also affected. One can see why the Competition in Contracting Act and similar legislative reforms are causing procurement paranoia in government acquisition circles. No one has yet computed the total effect of these laws on the costs and delays to major systems. There was obviously a lack of prior discussion and detailed analysis in Congress about the limitations and costs of "creating" competition.

As discussed earlier, an adequate analysis, planning, and simulation phase in the formulation of new laws like the Competition in Contracting Act would prevent or at least reduce the number of inconsistencies and impracticalities in procurement laws. A need for restraint in adding more laws is clearly indicated by this recent experience. Every new law or system of rules and regulations merely adds more complexity to an already complex maze of controls, restrictions, and wickets procurement officials must confront in trying to get anything done. Removal of flexibility in decision making has already slowed up the process in many areas because of the need to thoroughly check out and interpret all applicable statutes and regulations. Piling more regulation onto the already overloaded system tends to lead to inaction instead of expedited procurement. Judicious removal of counterproductive laws and regulation itself will require in-depth planning to avoid the same flaws that have been encountered in the past—such as inadequately conceived phase-in plans for new laws. A thorough cross-checking of present and past legislation could have surfaced potential conflicts and permitted the new law to be consistent with others still in effect. Any new legislation by Congress must consider the total impact of the legislation on both industry and government in carrying out the procurement process and should take into account long-term as well as short-term impacts. The cost and time impact of new laws must gain equal importance to their accomplishment of social, moral, or legal objectives in the overall assessment of the necessity and wisdom of legislation.

James L. Kington, editor of *The National Estimator* said, "Solutions to the problems that would provide for effective competition were not addressed before the legislation was issued.[12] Estimators are used to working with difficult sets of conditions, but the current environment is really a tough one." He pointed out that breaking procurement into smaller pieces with even more bidders adds to the already tremendous workload. The Air Force expects to have over 3 *million* transactions a year. With subdivided procurements, the government is going to have to decide how to treat the task of total systems integration and configuration management "where the prime contractor is no longer responsible for equipment procurement." The use of "breakout buys" and associate contractors disconnects firms that are supporting a project from direct contractual obligation to and control by a prime contractor. The National Estimating Society, in its 1985 Washington conference, developed a consensus that "we should not stop trying to make ourselves heard in relation to the task of solving those problems created by legislation."

In a 1986 educational conference of the National Contract Management Association,[13] retired Air Force Lieutenant General James Stansberry warned, "There is an inherent contradiction between increased competition and the enhancement (even the survivability) of the industrial base." Stephen Rowan, director of contracts for the Raytheon Corporation, agreed that competition for its own sake is not good and urged its proponents to "slow down, wait, and assess the effects of new competitive procedures on the primes and subs and try to solve current problems through *cooperative instead of adversary methods*" [emphasis added].

Industry has been able to foil the competition initiatives anyway. In the Navy's Osprey program, "when the only two technology leaders teamed, this negated any design competition strategy."[5] At last reading, the Navy expressed its intention to split up the team and force both companies to compete for shares of the full scale production contracts. The question here is whether the savings incurred by enforcing competition are justified by the high costs of dual sourcing. A perhaps inadvertent evasive action to competition took place at NASA's Goddard Space Flight Center on a part of the space station study project. General Electric and RCA, the two leading contenders for the work, became one when GE acquired RCA, creating a de facto sole-source situation. If the leading companies join together to form a team, then everyone gets a piece (albeit smaller) of the action. Forced splitting of the team into two or more entities, however, increases the costs of maintaining procurement options.

Lynn Bateman, writing in Government Computer News,[14] said:

> *Many bills put together by well-intentioned but poorly informed Congressional representatives end up costing the government money, and the general public is not adequately prepared nor sufficiently forewarned to handle their effects. The Competition in Contracting Act, for example, has cost the government a great deal of money in many instances when its original intent was ostensibly to cut costs.*

Those who have observed the effects of overregulation in government procurement conclude that it will take some time to clear up the maze of problems created by trying to legislate across-the-board, simplistic solutions. Most are in favor of reversing the current trend toward legislation as a cure.

A Call for a "Moratorium" on Procurement Legislation

There is a rising ground-swell in both industry and government for the Congress to impose a *moratorium on procurement legislation* until problems brought on by the current set of laws are solved and until a long-term, unified well-tested and well-thought-out procurement statute can be developed.

Thomas G. Pownall, chairman and chief executive officer of Martin Marietta Corporation, in a *Contract Management* article on improving defense procurement[15] said that Congress should:

> *Declare a moratorium on additional legislation pertaining to the conduct of the acquisition process until those regulations already legislated can be announced, understood, executed, and measured for effectiveness. While further new legislation would undoubtedly be well-intentioned, the greatest defense horror story of them all would be a rash of new laws that would decimate a system that has served our nation well throughout its history.*

Comments along this line are not restricted to industry. General Larry Skantze, commander of the Air Force Systems Command, told *Government*

Executive magazine: "We need a coherent approach to buying weapon systems. Piecemeal laws just cost too much. *A moratorium right now on procurement legislation would help*" [emphasis added].

Earle C. Williams of BDM[16] said, "As taxpayers and concerned citizens, we should oppose and speak out against the inappropriate and dangerously short-sighted efforts to procure research and development and professional services using techniques and policies designed to procure production quantities of weapon systems or standard tools."

In its search to find a magic law that cures everything, the Congress has been much too hasty in adopting across-the-board reforms that sound good but cannot be forced to apply in all situations. Already, lobbying groups have been formed *against* many of the procurement reform laws to advocate the repeal of those that affect special segments of industry. For example, The Association Group of Small Research, Engineering, and Technology Companies (TAG) has been formed in McLean, Virginia, specifically to suggest to their Congressional representatives relief for this group of companies from portions of the Competition in Contracting and other recent procurement laws. During the moratorium period, specific exceptions needed to maintain quality and excellence in science and research should be considered before enacting a simplified and enforceable procurement statute. If the exceptions are too numerous, one would wonder if a law is needed at all, except as an expression of public policy.

As we know from civil law, the existence of a law against something doesn't prevent it from happening; it only provides punishment if it does. Punishment may act as a deterrent if harsh enough, if violations can be detected, and if the law is enforced. Procurement legislators may be trying to eliminate all undesirable outcomes by imposing more laws, regulations, and controls when it is obvious that some people and organizations will continue to do the wrong thing out of ignorance or in hopes of not being caught or prosecuted even if penalties are severe. When laws become overly complicated or unreasonable and are difficult to enforce, even the honest and well-meaning will tend to slip into habits that disregard the law (take the 55-mile-per-hour speed limit, for example). Each new procurement law should be examined to determine if it truly presents a deterrent to abuses of the government's best interests. This includes an evaluation as to whether violations of the law can actually ever be detected and whether the law can be effectively (and economically) enforced. If one or both of these criteria are absent, then the law merely represents a statement of public policy. The enforcement cost must be weighed against the savings to be gained.

Procurement legislators have been trying to foster efficiency, economy, and effectiveness through intricate laws that are unwieldy, complicated, confusing, difficult to interpret, and tough to enforce. Means of detecting violations do not exist for some laws, and the penalties for violations are either inconsequential or nonexistent for others. Procurement laws or regulations that state "wherever possible," "whenever appropriate," "if feasible," "when deemed to be prudent," or "when in the best interests of the government" have no means of enforcement. The following modifiers are contained in the 1984 Competition in Contracting Act: "if time permits," "if the agency head determines," "to the maximum extent practicable," "with due regard to the nature of the property or services to be provided," "with reasonable promptness," "if it is likely," and "if it is necessary." The Congress and its staff recognize, at the time they pass laws, that such statements often reduce a law to a statement of Congressional policy. These clauses are intended to provide the executive branch sufficient flexibility to intelligently carry out the policies. Unfortunately, the executive branch agencies are, many times, unwilling to use the additional flexibility when regulations reflecting the laws are issued, causing such invalid interpretations as "competition at any price" and "award to the lowest bidder."

During a moratorium, the meaning and interpretation of the existing laws that contain these modifiers can be clarified and tested in the courts if necessary. Future laws enacted after in-depth analysis, simulation, and debate must avoid such ambiguities. What is the answer in the meantime? A wise industry executive once said[15]:

> *I don't believe that it requires a highly qualified body of regulations to encourage a trusted executive—be he in industry or in government—to recognize and do what is right. The weight of monolithic, overdetailed regulations does more to crush the exercise of individual discretion than any other factor. What we can do, and ought to do, is devote less effort to composing and enforcing shotgun procedures and prohibitions and commensurately more effort to obtaining, training, and retaining personnel blessed with the qualities that would obviate such regulatory harassment.*

Requirements Generation

In the view of those who have studied the major systems acquisition process, both inside and outside of government (and in *both* the executive and legislative branches of government), Congress must elevate its sights from the details of specific procurement reforms and specific major programs and systems to the processes of requirements generation and establishing program stability. In December 1972 a draft of the Commission on Government Procurement Report[7] containing a "Special Report on the Acquisition of Major Systems" was prepared and later submitted to the Congress in condensed form as part of the final report of the Commission.[17] The original draft of the report contained material that explained, in depth, the importance of the involvement of the Congress in establishing major goals, objectives, and needs, and proposed ways of getting the Congress involved in overall "proactive" planning and management: "The Commission believes an alternative role for Congress in system acquisitions is, at the outset, to debate the problems to be solved and their relative priorities and, from this, to review the fiscal limits within which solutions to our civil and defense problems can be created."

The report went on to describe how a yearly review would provide a more meaningful forum for the Congress to review overall programs and needs; how this review would provide the opportunity to establish an understanding between the executive and legislative branches of what goals are to be sought with each program; how it would remove the strong pressures to make premature commitments to a particular major system or project in order to gain funding approval; and how this better understanding would improve fiscal control over program commitments.

The presentation by the executive branch to the Congress of a set of national goals and objectives and a long-term dynamic national plan of major systems acquisitions to meet these goals and objectives, as recommended in Chapter 3, will provide the Congress with an overall framework to make decisions that will endure long enough to successfully see specific programs or distinct phases of programs through. In January 1987, the President sent a forty-one page document titled "National Security Strategy of the United States" to Congress along with the fiscal year 1988 and 1989 defense budget. This document represented strategy, yes, but not a plan. And it was restricted totally to defense. A more specific plan showing major systems and projects in both the civil and defense arenas should be developed with Congressional participation. The Congress, thus committed to a longer-range approach and view, can make more informed decisions on the funding of specific programs. This process, in itself, will help to foster program stability. The dynamic nature of the plan, coupled with longer range objectives, will permit smaller adjustments to meet shifting objectives as time passes. This process, then, will help to eliminate the wide swings in support for specific major systems throughout their lifetime. These swings, which have been a principal cause of cost growth, will then be reduced to mere minor adjustments to meet changing long-term requirements. One Congressional staffer wrote that this was tried in the fiscal year 1988 budget cycle, but that members voted their constituencies (i.e., they supported programs in their districts) rather than on national merit. If in prior years, trade-offs had been done earlier as part of a dynamic national planning process, these difficulties could have been overcome.

An Overriding Objective: Program Stability

The many studies of government procurement during the past several decades have varied widely in the identification of problems and in specific answers and recommendations. At various times the answer was thought to be "zero-based budgeting," "total package procurement," "fixed-price contracting," reorganization, baselining, work measurement, warranties, better education of the procurement work force, a civilian acquisition agency, legislated competition, truth in negotiations, different management techniques, enterprise programs, and so on. There is one area, however, in which virtually all studies agreed: There is a need for *greater program stability*. As has been shown, many of the studies attributed the lack of program stability directly to the Congress. It has become very clear, then, that the major contribution that can be made by Congress to effectiveness in system and project acquisition will be to *provide program stability*.

Colleen Preston, procurement counsel for the House Armed Services Committee, said:[11] "Budget stability will do more to control two of the three main variables affecting major systems acquisitions—cost and schedule [the other is performance]—than all of the management techniques one might devise."

In a paper titled "Perspective on Defense Systems Cost Growth" presented to the Operations Research Society of America in November 1984,[18] retired Lieutenant General Richard L. West, former Comptroller of the U.S. Army, listed five major reasons for cost growth in major systems. Second in importance only to "chronic overoptimism" (treated in Chapter 6) was "lack of program stability," described as follows:

> Lack of Program Stability. *This had become a fact of life caused by constant changes, including Congressional tinkering. Services [in the Department of Defense] were often faced with the problem of fitting the program to dollars; and the end result was program stretchout, the creation of major bow waves in the out years, and the necessity to program below economic quantities. It was acknowledged that constant program change was a poor way to manage and that program instability simply fed cost growth.*

On August 1, 1985, the Procurement Round Table, whose chairman is Elmer Staats, former Comptroller General of the United States and former member of the Commission on Government Procurement, chose "lack of stability" as one of six major problems in government procurement to be identified to the Packard Commission as study areas. On the same date, a National Security Industrial Association (NSIA) working group had come up with eight areas of concern, one of which was that the "budget cycle is too short for stability and economy." A subsequent working draft of prominent issues for discussion included "Program and Management Stability" as a top issue.

Everyone associated with the major systems acquisition process, both then and now, has been calling for increased stability in requirements, financing, and management. One government official told us that lack of stability causes major projects to be 30 percent to 50 percent more costly than they should be.

Perhaps the best study of the acquisition process we have seen in recent years is a study completed by the Air Force Systems Command on February 9, 1983, titled *The Affordable Acquisition Approach Study.*[19] Under the direction of Major General Melvin F. Chubb, the Air Force Systems Command Deputy Chief of Staff for Systems, the study was conducted to answer three fundamental questions: (1) Do programs take longer and cost more than they did previously? (2) If so, why? (3) What can be done to shorten the process or procure systems at a lower cost?

Research into these areas was guided by an executive steering group headed by retired Air Force Lieutenant General James Stewart and supported by the Analytic Sciences Corporation. The group performed a detailed review of *600* prior studies and reports dealing with the acquisition process, held extensive interviews, and developed a comprehensive perspective of the cost and schedule histories of 109 past Air Force major systems acquisitions. The final recommendation of this in-depth study was that

> . . . *the Air Force commit to an investment program which requires both a top-down and a bottom-up emphasis on* stability *if the Air Force is to achieve near optimum force modernization within available resources*

To help establish stability, the report recommended that the go-ahead decisions for full scale development and production be "tough gates." But that once these gates were approved, there must be a firm *commitment* of sufficient funding to carry out the approved program phase as planned. Once into production at an economical rate, "stay at this rate as long as there is a requirement for more systems." As the report says, a fixed production rate is "cheaper than stretching a program." Publishers of the report used a diagram on the cover which showed all of the initiatives recommended by the study leading into one final arrow labeled "*STABILITY*."

Does the Congress really have a commitment to completion target dates? To system performance achievement? And to cost ceilings? If not, then the Congress must be made a part of the hard-nosed decisions to proceed with major program phases to the extent that they are willing to continue to back the program financially throughout the phase's life cycle. At the same time, project managers must be rewarded for cost, schedule, and performance realism rather than optimism. Throughout the lifetime of a project, the Congressional committee membership, leadership, and controlling party can change one or more times. A committee that approves a "new start" may have a different makeup and political inclination at the time development is complete and a commitment must be made for production. These potential changes must be recognized and accounted for at the outset of programs. If these hard-nosed decision points are part of an executive branch dynamic national plan to be approved by the Congress, then the Congress can be a part of the solution rather than part of the problem.

Proactive Versus Reactive Leadership

Just as the project manager must employ proactive leadership to envision needs and to avoid problems before they happen, so the project manager's silent (or not-so-silent, as we have shown) partner must also exercise proactive leadership. In the past, the Congress has been generally a reactive body—reacting to problems each year as they occur and providing legislation to prevent the recurrence of major difficulties that have occurred. The reactive mode seeks to cure the disease by controlling the symptoms, while proactive methods are long-term preventative measures and durable, positive actions that overcome the root causes of the inadequacies and create environments and motivations for success. Proactive leadership by the Congress in major systems acquisition places the Congress in a position to approve national goals and objectives, and to make major system decisions that will endure throughout their ten- to fifteen-year life cycle.

The critical project success factors mentioned earlier in this chapter were summarized by Dr. Lewis R. Ireland of SWL Incorporated, in a presentation to the China International Conference Center for Science and Technology.[20] They were: involvement of the parent organization; comprehensive feasibility study; ongoing planning, coordination and review; commitment to the project completion dates; teamwork through clear objectives; commitment by parent organization to support the project; and clearly defined responsibilities and authority.

Critical success factors *within* the project were summarized as quality and depth of early planning, measurable technical progress, realistic schedule, clear decisionmaking authority, and maintaining design ahead of fabrication.

If project managers conscientiously fulfill the latter objectives, then, and only then, can the Congress faithfully fulfill its role as the parent or funding organization. Individual members of Congress, as well as Congress as a body, must, however, become accustomed to making the difficult tradeoffs involved in representing their constituencies and meeting national goals long before the authorization and appropriation process for each year in order to achieve stability in our major systems and projects.

The role of Congress is central in the acquisition of major systems and projects. Although Representatives and senators have terms of two and six years respectively, the study cited earlier shows that the average years of service of the average member of Congress equals or well exceeds the life cycle of most major systems. Knowing this, the objective we must pursue is to encourage the Congress, both individually and as a body, to adopt a longer range perspective—one that transcends their individual terms.

How can this be done? In a threefold manner: first, by agreeing on national goals and by providing the Congressional committees and subcommittees with the results of credible, in-depth, long-range planning that shows how specific systems and projects fit into an overall plan to meet these national goals; second, by increasing public awareness and knowledge of the benefits of continuity and stability; and third, by authorizing, appropriating, and committing funds on a multiyear basis for major programs. The initial agreement on national goals and plans must be done far in advance of the budgeting process itself, should involve both the Congress and the executive departments and agencies, and should be based on program lifetime costs rather than the costs of the next year, two years, or five years. Analysis should be done early and definition deepened until credible specifications and costs can be derived for each major system or project. Then, multiyear commitments of funds to be controlled by tenured project managers will be made available to successfully carry out each major system or project to its completion. A tall order? Yes. But these steps must be taken to provide the stability so desperately needed in our major systems and projects and to provide the most effective use of the billions of taxpayer dollars that are being spent for these undertakings.

Endnotes

1. Commission on Government Procurement: *Major Systems Acquisition—Final Report,* Commission on Government Procurement, Washington, DC, January 1972.
2. David I. Cleland and Harold Kerzner: *The Best Managed Projects,* Procurement Associates, Covina, CA, April 1986, GCS 8-86 D-1.
3. Interview with James Gregory, Director of Procurement (retired), Strategic Defense Command; Huntsville, AL, August, 1, 1986.
4. General Accounting Office: *DOD Acquisition Case Study Air Force Small Intercontinental Ballistic Missile,* General Accounting Office, Washington, DC, July 1986, NSIAD-86-45S-16.
5. General Accounting Office: *Strengthening the Capabilities of Key Personnel in Systems Acquisition,* General Accounting Office, Washington, DC, May 1986, NSIAD-86-45.
6. Gary Hart and William S. Lind: *America Can Win,* Adler & Adler, Bethesda, MD, 1986.
7. Commission on Government Procurement: *Special Report on Acquisition of Major Systems—Draft,* Commission on Government Procurement, Washington, DC, December 1972.
8. Roger H. Davidson and Walter J. Oleszek: *Congress and Its Members,* CQ Press, Washington, DC, 1985.
9. Roger H. Davidson and Thomas Kephart: *Indicators of Senate Activity and Workload,* Congressional Research Service, Washington, DC, June 1985, 85-133 S.
10. Roger H. Davidson and Thomas Kephart: *Indicators of House of Representatives Workload and Activity,* Congressional Research Service, Washington, DC, June 1985, 85-136 S.

11. Colleen A. Preston: "Congress and the Acquisition Process: Some Recommendations for Improvement," *National Contract Management Journal,* National Contract Management Association, McLean, VA, Summer 1986, Vol. 20, Issue 1.
12. National Estimating Society, *National Estimator,* National Estimating Society, Washington, DC, Fall 1985, Vol. 6, No. 3.
13. National Contract Management Association, *Contract Management,* National Contract Management Association, McLean, VA, February 1987, Vol. 27, Issue 2.
14. Lynn Bateman: "Factors Conspire Against Shorter Procurement Cycle," *Government Computer News,* Government Computer News, Silver Spring, MD, September 26, 1986.
15. Thomas G. Pownall: "Improving Defense Procurement," *Contract Management,* National Contract Management Association, McLean, VA, August 1986, Vol. 26, Issue 8.
16. National Contract Management Association, *Contract Management,* National Contract Management Association, McLean, VA, December 1986, Vol. 26, Issue 12.
17. Commission on Government Procurement: *Summary of the Report of the Commission on Government Procurement,* Commission on Government Procurement, Washington, DC, December 1972.
18. Gerald R. McNichols (Editor): *Cost Analysis,* Operations Research Society of America, Arlington, VA, 1984.
19. Headquarters, Air Force Systems Command: *The Affordable Acquisition Approach Study—Executive Summary,* Headquarters, Air Force Systems Command, Andrews AFB, MD, February 1983.
20. Lewis R. Ireland: *Project Management: Critical Success Factors and Keys to Effectiveness,* SWL, Inc., McLean, VA, August 1986, 58-07-86.

CHAPTER EIGHT

A Framework for Action

As can be seen in the foregoing chapters, the difficulty in consistently acquiring economical, timely, and effective major systems and projects for the government has not been in identifying the problems or even in finding suggested short term solutions. The difficulty has been in formulating integrated solutions that both warrant and can garner strong and sustained Congressional and public support. We do not presume to propose detailed solutions in this one small volume—only a framework on which long-term reforms can be built throughout the 1990s and beyond. Detailed methods to improve the process must be based on the same meticulous, systematic, in-depth analysis, planning, simulation and testing that have been recommended for major systems and projects themselves. This planning must avoid piecemeal solutions for the sake of expediency. Reforms to correct past deficiencies and to reach heretofore unattainable levels of excellence should be based on an integrated framework of sound management philosophies and major goals that can both spark the public's imagination in the short term and hold the public's interest over the long term.

We believe that an integrated, sustained approach to implementing the recommendations in Chapters 3 through 6 can be beneficial to all parties engaged in major government acquisitions. Industry will benefit from greater long-term visibility and less regulation; project managers will benefit by having greater tenure and improved program stability; financial managers will have the tools required to accurately track, control, and estimate costs; and the taxpayer will benefit by obtaining more affordable systems and projects. To ensure prompt and lasting reform, these key benefits must continue to be made known to those who will benefit.

The highest priority elements in this framework—ones which must be emplaced before the others—are commercial-type accounting, cost control, and estimating systems for the Federal government. In a two-volume report on *Managing the Cost of Government* issued in 1985,[1] the General Accounting Office lists four needed elements: strengthened accounting, auditing, and reporting; improved planning and programming; a streamlined budget process; and systematic measurement of financial performance. Notice that the GAO recommended strengthened auditing, not more auditing; and systematic measurement of financial performance, not more detail in financial reports. Although reorganization itself without major changes in management philosophy will not solve deep-seated problems, some reorganization will undoubtedly result as new standard financial approaches and systems are adopted for government use. The GAO lists system development efforts, organizational changes, and investments in people as required; but they add that much of the effort can be accomplished with little additional cost by coordinating new and existing financial system development activities, while systematically eliminating the old.

Some of the major financial advisors in the United States have chimed in to push for an accounting system in government and are standing ready to help implement the required changes. Arthur Andersen and Company has conducted studies over the past several years, resulting in some very specific recommendations on how to upgrade and validate government financial systems. This firm published reports on *Sound Financial Reporting in the U.S. Government*[2] and *Guide for Studying and Evaluating Internal Controls in the Federal Government*[3], in February 1986 and August 1986 respectively, which give both broad guidance and detailed instructions on how Federal accounting for major systems and projects can be improved. The firm points out in these documents that the government does not follow the financial rules it imposes on others; financial management functions are fragmented; the government has ignored its own laws (legislation passed in 1950 as a result of the Hoover Commission supplemented by related Treasury Department regulations) requiring an accrual accounting system; there is no requirement to publish financial statements; and programs are adopted without setting aside adequate funds. Why, then, has the public not responded in shock to these accusations and demanded that corrections be made? Perhaps it is public apathy, failure of the Congress to respond to well-documented needs, a seeming absence of interest by the press, or the lack of a clear way to show the disastrous effects of working without an adequate accounting system. The firm of Price Waterhouse had also done an extensive study on government accountability and published its results in a blue-covered report in 1983 titled *Enhancing Government Accountability*[4] and in other reports. Both of these firms have worked with several of the larger state governments and have found that accurate, credible accounting and financial reporting is possible using advanced state-of-the-art automated systems. These two firms, as well as other firms and professional accounting organizations, are still asking why the federal government cannot become well organized financially.

As shown earlier in the book, the ability to know where we are financially is an essential element to knowing where we are going in our major systems and projects. We as taxpayers and citizens must continue to convey our concern about the financial health of our country and our ability to compete with other countries. The desperate need for a financial and accounting initiative alone should be enough to force a public outcry for a better organized government to meet our major defense and civil needs efficiently. When combined with the other three major areas of potential improvement, however, we believe that there is a convincing basis for a dedicated effort to redirect the present course of events and to go after the potential synergistic benefits that can be derived from moving forward on all required fronts to improve our way of obtaining big results from big systems and projects without further moving our country toward a financial crisis.

The improvement of accounting, estimating, and financial systems and methods by itself, however, offers little excitement to the average citizen. To many people, the sums involved in major government systems and projects are so enormous as to be beyond comprehension. Unfortunately, among many citizens there is a feeling of apathy and helplessness about ever really improving the cost efficiency of big government programs. The federal budget is viewed by many as a big "pie" with too many people trying to get their portion. Something more than mere "fiscal responsibility" is needed to truly capture the imagination, interest, and support of the public on a sustained basis to force major improvements.

The one key element of reform that could have more potential for instilling a deep and sustained interest in long-term efficiency in major acquisitions is a specific plan—a dynamic long-range national plan—that will visibly depict systems and projects over the next ten to fifteen years, and how these systems and projects fit into current and past efforts. One such integrated program plan that has already captured the imagination of many is the Strategic Defense Initiative, or "Star Wars" program. This program has had much the same effect on the general public as the Apollo moon landing program did in the 1960s. Its appeal is based partly on the high-tech nature of its potential systems, partly on the fact that it is a defensive network rather than an offensive weapon, partly on the fact that it is long-term rather than short-term in nature, and partly on the fact that it has many potential technological benefits. Those who are involved in the program already cannot help but experience a sense of excitement and anticipation about what may be coming. A dynamic, long-range plan for all of our major systems would include programs such as SDI and other defense systems, and many others, such as high-tech communications networks, transportation systems, and energy sources. The public's awareness of the need for effective and efficient major systems and projects could be enhanced beyond measure by the advancement of bold new long-range programs showing actual goals, objectives, milestones, and benefits. The dynamic national plan for specific systems and projects could and should include credible and clearly stated advantages; potential technology transfer benefits to commercial products, services, and processes; and potential economic benefits to the localities, industries, educational institutions, and citizens throughout the United States. The plan must be courageous; imaginative; potentially fruitful in multiple disciplines, industries, and professions; and presented in an easily understood and interpreted way.

The dynamic, long-range national plan for major systems and projects will provide a rallying point for those who want both to see and to influence where we are going in the next fifteen years. It undoubtedly will be a subject of much discussion in political campaigns, speeches, lobbying, debates, and discussion in the subcommittees and committees and on the floor of the Congress. It must be visible and alive. It must provide for increasing stability in approved programs, yet good flexibility in dynamically adjusting to new needs as they arise by phasing in new programs as they can be afforded and phasing out old programs as their missions have been completed. The absence, to date, of an active and responsive long-range national plan, we contend, has been a principal reason for the apparent inability of reformers to make much progress in revamping and upgrading the major systems acquisition process. Perhaps some have feared the implications of having such a plan at all because of apprehension that it would become too rigid, inflexible, and unresponsive. And others may resist systematic and methodical long-range planning because of their fear that a "big brother" government with such a plan would impose its will on the people. With our current system of cross-checks, checks and balances, and safeguards, there is little chance that a plan mandated to be a flexible guideline could become a rigid national obsession that could start controlling people's lives and companies' decisions. Instead, we believe, a progressive, imaginative, positive, far-reaching plan of action could ignite initial enthusiasm and foster continued interest in the national acquisition process. This plan, designed to be a major motivator to achieve sustained public support, joined with improved financial accountability, tenured project managers, and a selectively deregulated procurement process, could comprise the national rallying point required to make progress now that has been heretofore elusive.

Because of the unique relationships between a project manager, his or her project, the hierarchy of the executive branch, and the Congress, stabilizing the tours of project managers to match the timing of major programs or their principal phases is essential. Establishing credibility in cost, schedule, and performance estimates is a two-way street between the project's management and the Congress. Despite the fairly good stability of personnel in Congressional staffs and in Congressional representation on committees, there still needs to be someone in government—some individual—who can be held responsible to the people on a longer term basis for making realistic estimates and for carrying out programs in a cost-effective manner. This one person has to be the project manager. We believe that the same person who is going to be charged with doing the job should have the task of making time, cost, and performance estimates. And this same individual must have the freedom, motivation, and organizational clout to "tell it like it is" if potential cost or performance problems are indicated before the project is started. The project manager must have sufficient experience, tenure, and organizational support from the top down to exhaustively search out, reveal, and then correct problem areas before the project gets going and at each phase along the way. He or she must be assured of job stability and promotion potential, not based on starting more projects but on completing projects within original estimates.

Far better than providing another artificially induced independent check or perspective from outside the agency or department is the reliance on an experienced tenured project manager who has the desire, authority, and charter to call upon peer experts from wherever they exist, who will provide "independent" inputs directly to the project manager during the decisionmaking process—which is where and when they will do the most good. Air Force managers must feel free to call upon NASA managers for peer reviews in space matters, for example. Nuclear weapons managers must feel free to—and have the resources and organizational backing to—call upon nuclear energy experts, consulting physicists, industrial counterparts, the Department of Energy, and the academic community to provide peer inputs to important program decisions. Our project managers, because of their unique relationships with other organizations and their control over enormous resources, must come from an elite corps of highly skilled business/technical administrators and communicators. They must have proven their abilities in setting and meeting commitments through a progression from smaller programs to larger ones. They should not only be highly skilled and trained in project management techniques but must have the type of personality, attitude, and drive that is required to get the job done on time and within the allocated resources. Obviously, these are not individuals who have had jobs doing this and that for a number of years, but they must be specifically educated, trained, screened, subjected to duty on live and active programs, just as astronauts are finely tuned for space travel. Major systems developments and civil projects are inappropriate places to serve as training grounds for inexperienced or untrained project managers: these programs have far too great an impact on the economy, our national welfare and security, the well-being of private industry, and our international position in competing for resources. We know that this breed of project manager exists because we have seen them at work in real projects: excellence achieved in many other projects attests to the existence of people who can make commitments and stick with their projects.

The other side of the two-way street is to achieve a relationship between the project manager and his or her own agency and with Congress: a relationship that will foster confidence in program stability throughout the project. Congress, as well as the advocating service, department, or agency, must resist the temptations of competitive optimism that tend to permit projects and systems to start out inadequately planned, defined, and funded. Congress must avoid the blandishment of wielding the money whip at the last moment in affecting crucial program decisions. Thus, the Congress itself—involved in the project since its inception and prepared for long-term support—can help prevent the devastating and disruptive effect of funding reductions on the morale of those in the program and on the economy and efficiency of the program itself. It was the viewpoint of one Congressional staffer that "Congress generally changes a program only when it is already in trouble because it is behind schedule or over cost. For example, the B-1 bomber, the Bradley fighting vehicle, the MX missile (schedule slippage though Congress committed to at least 100 units)." Congressional involvement early in the overall planning process can help provide much preventative medicine *before difficulties arise*. As the parent or funding organization, the Congress must provide stability and consistency of support for the major program or its phase, just as the project manager must provide the necessary continuity and consistency of management control. Despite recent laws and regulations to the contrary, many project managers are still changed or transferred in the middle of their programs or in the middle of major program phases to other assignments. This continues because of a lack of appreciation for the detrimental effects of changing project managers in midstream, and by a need for the public and the Congress to recognize and to require project management tenure as an enforceable policy.

As in the other structural members of this framework for action, it is hard to justify strong and sustained action on this recommendation alone because it relies so heavily on the other structural elements of a modernized acquisition philosophy. For example, there would be little reason to insist on long-term project manager assignments if financing continues to be unstable. When funding is radically cut back or specifications changed, it's a whole new ball game. At this point, agencies often bring in a new individual and have him or her formulate a revised program to meet changed requirements. Project managers in a truly reformed acquisition process will need precise, current, accurate accounting and financial management data using the latest state of the art in fund estimating, tracking, and control. Hence, reforms must be coordinated to permit stepwise adoption of improvements when they can be synchronized with improvements in integrally related areas.

A part of the training of potential project managers should be the review of past programs and an analysis of the reasons for success or failure in programs. Project managers will need to know *why* they are following certain procedures—*why* these procedures have worked before—and *why* they may or may not work in a given situation. They must have a firm grounding in modern conceptual design, preliminary design, systems engineering, analysis, simulation, and testing. They must understand the orderly progression of a program from its embryonic stage to its final production, operation, and support phases; and they must understand that definition must deepen with each phase. They must be able to distinguish between the need for sparse specifications and requirements early in programs to enhance innovation, and the need for a transition to very detailed specifications and requirements in the final production and operation phases. And they must have the knowledge of when to freeze design or say "no" when unproven changes are proposed that will unduly affect costs or schedule.

The final structural element in the framework of a new, more efficient and more effective major system and project acquisition process is simplification and time-phased, selective deregulation. First, a moratorium must be placed on further procurement legislation until a detailed, integrated, systematic plan can be derived to synchronize procurement law with the other initiatives on national planning, financial improvement, and project management tenure. The Congress itself, assisted by the subcommittee and committee structure, the Congressional Research Service, the Congressional Budget Office, and the General Accounting Office should develop a phased legislative plan that will provide a statutory foundation for a multiyear program to simplify and improve the major system and project acquisition process. Current laws should be carefully reviewed for areas where they can be eliminated or combined. Outmoded laws should be rescinded or replaced with updated statutes. Plans should be made for a time-phased transition away from the present method of using government procurement to enforce socioeconomic objectives. Socioeconomic objectives which have long since been achieved should be terminated, and others which still need further emphasis should be phased-in as civil laws and gradually excluded from the procurement law arena.

With the assistance, where required, of the President, the Congress should be prepared to enter into a cycle of legislative planning that will transcend political, geographical, and institutional boundaries with the goal of developing a transition plan to a new environment for excellence. This project, in itself, should be a major national goal—to get our business in order to handle the ever-increasing budgets and resources that will be required of the major systems and projects of the next century.

The Role of the President

Throughout this book, we have said little about the role of the President in major systems and project acquisition. Just what is now and what should be the role of the President in the new environment for major systems and projects?

Several scholars[5–7] have observed the emergence in recent years of what is called the "Administrative Presidency," in which the President assumes the role of manager of the executive branch as well as the goal-setter for national priorities. This is occurring in spite of the fact that the Congress, as early as 1789, gave statutory powers directly to agencies, and that Supreme Court decisions in 1838 and 1935 upheld this doctrine. But the public and the press continue to hold the President responsible for activities in the executive branch and are increasingly demanding a "hands-on" Presidency in which he or she plays an active role in the management of the country. In this emerging role, the President will have an increasing responsibility for major acquisitions, and, in our view, will be taking a more active role in planning as well as in reviewing and managing the sequencing, approval, and overall conduct of major programs. Experience has shown that big projects need sustained proactive management from the top down in the executive branch, meaning from the President.

The problem of real-time Presidential management of our mix of major systems and projects, as pointed out by Stephen Hess in *Organizing the Presidency*,[5] is that "Presidents have assumed too many roles and have been woefully miscast in one—that of a manager." Many Presidents have not had the experience, training, or even the desire to be a manager. Hess suggests that the President, then, must rely heavily on someone else to help carry out the management of the executive branch. He recommends a "collegial Presidency" in which the cabinet members take a larger management role, and that the director of the Office of Management and Budget, the super-manager, be elevated to a Cabinet position. Hess paints a picture of an active, involved President. He recommends increasing the frequency and regularity of Cabinet meetings and proposes semiannual "mind-stretching exercises and presentations that have practical application, or that identify long-range concerns." Parts of these Cabinet meetings could include attention to long-range goals and to major systems acquisitions to meet these goals. Recent history has shown that a detached management style and too much delegation has its definite drawbacks, and that the President is responsible for what goes on regardless of whether "he knew" or not. *Initiatives in our major systems acquisition process, then, should not only strive to keep the Congress informed, but should assure that the President himself is informed and involved and has an opportunity to participate in the selection and carrying out of our major systems development programs and civil projects.* Whether or not it is elevated to a Cabinet post, the Office of Management and Budget, as the management arm of the President, could synthesize national goals; develop and maintain a dynamic long-range plan and budget for major systems acquisition in keeping with these goals; and present the plan to the President and his Cabinet, and subsequently to Congress, for approval.

History has shown that work can be decentralized, but that ultimate management responsibility can not. "The buck stops here" still holds true. The Office of Management and Budget is uniquely suited to integrate multiyear budgets into long-range programs. After gaining years of experience in preparing the mammoth federal budget and tracking it through many perturbations as it makes its way through the Congress for final approval, the OMB has developed a number of effective techniques accompanied by computer-based systems for automated budget preparation and tracking. OMB now has the capability to receive electronic submissions of budget numbers directly from the departments and agencies in the federal government. When agencies submitted their proposed 1988 budgets, more than one third of them were delivered electronically rather than on paper, and the electronic submissions represented 90 percent of all spending. OMB has two large computer programs on its mainframe IBM 3083: a Budget Preparation System (BPI) and a Central Budget Management System (CBMS). The BPI produces the budget, and the CBMS tracks changes to the budget that are made throughout the year. These same systems can be used for the multiyear budgets that accompany long-range plans. A number of states have similar budgeting systems that are "megasystems" in themselves, having taken many years and many dollars for development. *With these highly capable computer systems at hand, and with many more such systems becoming available on a regular basis, the OMB can correlate requests from agencies with income projections from the Treasury Department to develop realistic budgets that fit into long-range plans, and they can rapidly determine if and where expenditure projections exceed income potential.*

A top level, long-range planning activity in the executive branch tied in with budget preparation could result in sizeable cost savings by identifying joint programs that could benefit several departments, agencies, or services. So far, joint major systems acquisition by the military services has been an "elusive strategy."[8] The General Accounting Office concluded, among other things, that there was an "unwillingness by the services to make joint programs work," to which the Department of Defense responded: "Nevertheless, the services, though willing, may still not be able to resolve all requirements disputes, and there is no super-service military umpire to have the final word and make it stick." Since he who controls the budget controls the program, the President and the Congress should be able to make joint programs stick. Furthermore, there are several examples of very successful joint programs now in progress which prove that they are feasible. The MILSTAR satellite communications system[9] is one. Even more appropriate for joint programs are the large budgeting, estimating, data base, communications, and information systems now being planned and/or developed by the various federal agencies. The many common areas involved in managing a large federal agency that require major automated systems could be coordinated to permit shared or distributed use of common or generic information systems, software, and data bases.

There is not a successful company today that does not have some sort of long-term goal or plan that tells where the company wants to be, say, ten or twenty years from now. But the *U.S. government, which sponsors the development of megasystems and megaprojects that have lifetimes of ten to twenty years or, more does not now have an integrated strategic plan.* This is why the final presentation of the Major Systems Acquisition Study Group to the Procurement Commission members in 1971[10] included Part 4a: *PLANNING*. The recommendation stated that a national planning function and process be provided and that it be the President's responsibility. The recommendation stated further that its purpose was to: (1) provide continuous long-term planning for selection of national goals, ordering of priorities, and use of resources; (2) integrate planning of major systems; and (3) improve public and Congressional understanding. The recommendation came after two years of study, during which time leaders within both industry and government told the Study Group that this was a major need and that it would have a positive and constructive influence in reducing costs by stabilizing major programs.

Problems in achieving this type of planning have resulted from one of the very foundations of our democratic system: that it includes separation of powers and a chief executive with limits on tenure. The question, then, is: Can a President effectively carry on the programs of a predecessor? Experience in successful programs like sending a man to the moon has shown that it is possible to garner long-term, sustained public support for major programs that span more than one administration. Although it may be feasible to prepare a national plan and budget for major systems, it must incorporate and take into account agency mission plans, goals, and budgets and must retain agency accountability for programs. The national plan is nothing more or less than an integrated, high visibility mechanism for projects that impose long-term obligations of enormous magnitude on the Federal budget. It must be developed as an integral part of the current planning process rather than as a new system, organization, or procedure.

This, and the other recommendations that have resulted from the myriad studies that are synthesized in this framework for action, can be done with existing organizations and without the establishment of new offices, czars, hierarchies, or agencies. As discussed in Chapter 1, reorganizations and new organizations by themselves do not produce success. We only need to put current organizations to work doing those things that will result in long-term benefits in an already too-much-studied and too-little-improved arena.

Evolutionary, Phased Development

Evolutionary development based on the well proven and tested sequential phased acquisition approach is clearly a key element in any long-term effort to improve the major systems acquisition process. Time and time again a phantom culprit has been named as the guilty party in causing programs to fail to meet cost, schedule, and performance goals. The name of this culprit? *Uncertainty.*

Uncertainty breeds unknowns, which also get a share of blame for program failures. Devotees of the mysticism of uncertainty have even subdivided unknowns into "known-unknowns" and "unknown-unknowns." If uncertainty and its family of unknowns are the problem, then the solution to the problem is to eliminate uncertainty and unknowns. We have shown, in describing tools available to project managers, that there are many effective "uncertainty removers." These lie in the disciplines of simulation, analysis, testing, evaluation, quality assurance, and systems engineering. Detailed definition based on sound engineering development practices will remove uncertainty. Top level officials at NASA have told us informally, for example, that the tragic Apollo fire on January 27, 1967 that killed three astronauts, as well as the space shuttle *Challenger* accident nineteen years later on January 28, 1986, which killed its crew of seven, could have been avoided if the warnings of highly qualified engineers had been heeded regarding the use of realistic environments, extensive testing with appropriate redesign, and meticulous quality control.[11]

Just what is evolutionary phased development, and why does it have the potential to stabilize programs? Evolutionary, phased development recognizes first that development must be completed before production can start. It is the antithesis of *concurrency*, in which an attempt is made *to develop an item and produce it at the same time*. (Publilius Syrus wrote, "To do two things at once is to do neither.") Evolutionary, phased development requires sequential application of engineering disciplines but permits the introduction of improvements during a production program, either through block changes to be incorporated in future production lots or through retrofit programs. The secret behind successful evolutionary development is to *subject the changes or modifications to the same rigorous testing and verification experienced by the parent system*. This rigorous testing and verification, in order to be fully valid, must include *re-analysis or retesting of the whole system with the changes or modifications included*. Evolutionary development is, then, a methodical approach to first developing the parent or baseline system before production starts and then upgrading it in increments to new baselines only after undergoing the same exacting verification process. When unproven modifications are not interjected into production, the detrimental effects of "concurrency" will be avoided. Evolutionary development can proceed hand-in-hand with dynamic national major systems planning and acquisition because both provide enough flexibility to accommodate time for removal of remaining uncertainties as well as for injecting beneficial changes when needed.

Documented examples have shown that the best way to deal with uncertainty in major programs is to substantially reduce or eliminate it through exhaustive and meticulous analysis, simulation, testing, and evaluation prior to commitment to subsequent steps in development or production. Throughout the past several decades, engineering disciplines have emerged that are uniquely suited to the removal of uncertainty from large programs. These engineering disciplines are based on the use of high-technology tools and methods to do a systematic analysis of the performance, interfaces, quality, reliability, and potential cost of major systems. Mainframe, mini, and microcomputers are playing an increasingly important role in this analysis because computers are accelerating the capability to carry out these engineering disciplines with great speed and accuracy. Engineers are now able to explore thousands of design concepts under thousands of different operating conditions without building the first piece of hardware. Once prototype hardware is built and tested, modern computer instrumentation and test analysis techniques provide a phenomenally fast feedback to permit rapid redesign and retesting of upgraded parts, subsystems, and systems prior to baselining production designs. Computer-aided design can now feed directly into computer-aided manufacturing to speed up the design–testing–redesign–retesting cycle.

Airplane designs can be "flown" on computers that simulate intricate aerodynamic forces on all parts of the aircraft. Thermal, vibration, shock, and pressure effects on electronic, electrical, and mechanical parts of major systems can be accurately depicted in sophisticated three-dimensional computer programs before the parts and systems are even built in the laboratory, model shop, or factory. Stress corrosion, interfaces between major subsystems, and electronic circuit performance can be simulated under almost any condition to permit a fine-tuning of design that was never before possible. Computer-driven machines can take designs from computer-generated "drawings" and manufacture parts without ever committing their design to paper. Pilots can fly aircraft before they are even built through the use of computer-controlled visual simulators that react precisely the way a real aircraft would in an emergency situation. The pilot of an aircraft or driver of a tank can explore the full limits of the vehicle's performance under any conceivable scenario without fear of encountering a life-threatening situation or of a catastrophic "crash." Astronauts are able to dock simulated spacecraft with simulated space stations in an environment so realistic that blood pressure and heartbeat measurements go up and down just as they would during real spaceflight.

Through the use of the latest high speed, high-tech computers, configuration engineers can test the interactions between complex electronic and mechanical devices; cost estimators can develop credible resource requirements for engineering and manufacturing of parts and systems; and managers can track cost, schedule, and performance using sophisticated critical path networks and bar charts. The increasing speed and increasing economy of computer power make it possible to design, test, estimate, and recycle the design "on paper" (meaning on the computer) many, many times to fully explore interactions between cost, schedule, and performance. Sophisticated life-cycle cost engineering and design-to-cost engineering techniques and tools are becoming available that explore program limits, test the influence of widely diverse constraints on programs, and determine impacts of changes. *Systems Engineering and Analysis*, by Blanchard and Fabrycky,[12] and a new version of the book (1989), provide fascinating insights into these emerging disciplines in systems engineering.

As all of these high-tech tools and methods are helping to remove uncertainty from the "front-end" of the acquisition process, new testing, manufacturing, and evaluating techniques, also aided by sophisticated instrumentation, electronics, and computer analysis, are speeding up the process from design to actual product or system development. The results of scale model, prototype, or initial operational unit tests can be fed back into design programs rapidly to update and fine-tune prototypes for low-cost and optimum performance.

The use of these now well-developed and sound engineering principles, provided that they are applied meticulously and conscientiously, can remove uncertainty in programs *prior to* entering into firm contractual agreements, thereby virtually eliminating the possibilities of change-driven cost growth. Experience has proven that cost growth *can* be stemmed if the work is defined and specified in sufficient detail, and if excellent management principles are applied. Where all remaining uncertainty cannot be eliminated through meticulous definition or by reducing the scope of contracts to cover only definable work, design-to-cost engineering principles can be used to adjust performance, quantities, manpower, overhead, labor rates, or schedules to stay within cost ceilings.

These tools, when made available to the project manager and contracting officer, if coupled with sustained resource commitments and excellent accounting systems, permit greater definition, reduced uncertainty, and more stable programs. These engineering disciplines permit more detailed definition of scopes of work and resource limits before government agencies and their industry counterparts commit to a contract and can be the basis for evolution to fixed-cost, phased procurement for major systems.

We Propose Life-Cycle Management

We can define life-cycle management as a management philosophy intended to foster continuity and a sustained dedication to excellence in meeting program and system objectives in keeping with national goals. Life-cycle management includes continuity of management personnel, principles, practices, and procedures at all levels and at all steps in the acquisition process throughout the lifetime of the project. *It includes continuity in legislation that applies to the program and calls for restraint in changing program philosophies in midstream.* It suggests that a program initiated under one set of procurement laws should be carried through to completion under those same laws, that project managers should be permitted to stay on the job and still receive merit promotions, that funding objectives and goals should remain stable, and that disruptive external influences should be prevented during each major program phase.

As discussed in Chapter 5 and again earlier in this chapter, significant changes are needed in the present methods of training, assigning, motivating, and promoting project managers and their team members to achieve project stability and to move toward life-cycle management. According to one recently retired project manager, military and civil service personnel policies are not flexible enough to permit long-term assignment of project managers and other key project personnel. Grade levels are attached to specific positions. This prevents retention by the project of those who want to continue to move up the ladder. These policies must be changed in order to implement life-cycle management.

Most excellent managers are not willing to wait for the completion of a full program phase before being promoted. An example was given in Chapter 7 of a project management position that was upgraded to general while occupied by a full colonel, resulting in reassignment and the loss of a good manager. Good project managers can literally be driven out of their position by restructuring of grade levels during the project. Life-cycle management will require enough flexibility to promote excellent project managers in their present positions in order to retain their vitally important skills. Grade structures should be made sufficiently flexible to permit placement of the best available military or civilian project manager who can remain with the program, regardless of grade or rank, and then to permit within-the-job promotions as a reward for excellent sustained performance. Project engineers and contracting officers, as well as project managers, should also be groomed for permanent, progressively responsible positions.

Periodic changing of the project manager and project team personnel may be good for the careers of the people, but it is very hard on the project, and it does not support the concept of life-cycle management. To adhere to the recommendation that the project manager be a "proactive" leader, he or she must be involved in the day-to-day operation of the project. To relegate this day-to-day leadership to a deputy and to assign the project manager to cover "external interfaces" is to absolve the project manager from responsibility for those things that go on within the project. Projects have been successful because they have had strong, sustained, proactive leadership. If this leadership can only come from the deputy, then he or she should be promoted to project manager, and the project manager should be promoted to senior advisor. But even the role of senior advisor requires continuity, sustained interest, and intimate contact with the program. Major programs can no longer merely serve as management training grounds for career personnel who spend one to two years with each program and then go on to a completely different assignment. The mistakes that can be made in passing the baton are too great, and the opportunities to change project objectives or schedules for the sake of change or to fail to honor previous commitments are too numerous. Assigning a new manager to a troubled program merely changes the person in charge; it does not, in itself, solve the problems. It merely absolves the previous manager from the responsibility and allows the new one to restructure the program by reducing operational capability, slipping schedules, or increasing costs.

A management policy of retention of responsibility and authority by competent, experienced project managers from the outset of programs will result in greater forethought, planning, and analysis before making commitments. *Failure to meet commitments cannot be taken too lightly if meeting commitments is to become a way of life.* To quote Publilius Syrus again: "Pardon one offense and you encourage the commission of many." Project managers need to participate in and approve major program commitments and then be around to see that these commitments are carried out. They should be held accountable for their failures, as well as rewarded for their successes. Life-cycle management would foster less emphasis on the initial price of a proposed system and much greater emphasis on investments in competing candidate systems, with final selection based upon relative performance and relative total cost and affordability to acquire and operate the system. Life-cycle management requires excellence in accounting and estimating systems, in harnessing funding instability, and in producing a single, durable procurement statute.

Although better accounting and estimating systems in themselves are not expected to solve any major problems in the major systems acquisition process, their effective *use* will give greater visibility to problems before it is too late to do anything about them. The first steps have been taken toward a standardized, government-wide accounting system that would track actual expenditures against agency budgets to enable federal managers at all levels to see exactly where they stand financially and where they are going. Well-established principles of accounting that have long been used in private industry are now being adopted within the government, and automated systems using the latest computer and software technology are becoming available to handle the tremendous daily number of government transactions. But we still have a long way to go. Two major needs still exist to credibly connect actual expenditures with budgets and to form a basis for realistic financial projections: (1) a system for retaining longer term financial data to be used to project future monetary needs; and (2) compatible and uniform accounting and estimating systems.

If federal agencies are not able to collect comprehensive, long-term historical cost data or to determine where they have gone wrong in their estimating, they will have little opportunity to use this information for improving their estimating on future programs. The General Accounting Office, in reporting an incredible 263 percent cost growth (*above* the initial estimate) in the Federal Aviation Administration's Airport Surveillance Radar (ASR) upgrade project,[13] for example, included an even more incredible fact as a footnote to "Reasons for Cost Growth." The GAO report said: "Because of many changes in the ASR replacement program and the lack of information for tracking these changes from one estimate to another, we are not able to identify or estimate the amount of cost growth for the components of the ASR system." The footnote to this statement said: "*Certain requested information had been discarded because of the agency's policy of retaining information for only 3 years*" [emphasis added]. This report was sent to the heads of the Senate and House Budget Committees but required no response. It listed no required actions and contained no provision for increasing the time of data retention.

Longer term data is provided for most Department of Defense programs through the Selected Acquisition Reports (SARs) for major systems. All agencies, including DOD, should adopt the GAO specified format for their major programs, which requires visible tracking of costs and schedules versus initial estimates. Cost and budget estimators need traceability of costs in order to realistically project future costs of systems. Just as every agency has had its own accounting system in the past, each agency has had its own estimating systems and techniques. Estimating systems need to have structures and coding that are compatible with accounting systems to track actual expenditures against estimates. Compatibility of estimating and accounting systems between agencies will decrease costs because generic systems can be used by multiple agencies, and government-wide compatability will permit meaningful comparisons and more effective use of actual data for estimating the costs of future major systems. As proposed in Chapters 1 and 4, the most urgent of all national needs is a reliable accounting system for the federal government, one that will match the quality of those now required of and existent in the best of private industries. Close on the heels of this priority and integrally associated with it is the need for a *uniform estimating system*.

There is no shortage of criteria for standardization. The 1987 *Cost Estimator's Reference Manual*[14] contains a bibliographical reference of twenty-two recently published books that spell out the steps, disciplines, techniques, methods, and tools available to the cost estimator for preparing credible cost estimates. Many more books and periodicals on the subject are appearing regularly, and several professional societies have been established to promote better cost estimating both inside and outside of government.[a] Several agencies have their own extensive libraries of cost data, references, and computer programs that are available for use by other federal agencies. As soon as unified systems and criteria are developed, a federal accounting and estimating handbook could be issued. This handbook could be used as a guide for the integration of estimating and accounting systems and for the enhancement of uniformity, consistency, and credibility of accounting and cost estimating on a government-wide basis. The government could then begin adopting and using uniform, simplified, government-wide accounting and estimating systems to forecast the resources required to ac-

[a] The National Estimating Society, Institute of Cost Analysis, The American Association of Cost Engineers, and International Society of Parametric Analysts.

quire its major systems and to enhance the credibility and accuracy of the life-cycle financial management of programs.

Life-cycle management will both enhance and require program stability. Congress itself can make major strides toward program stability by authorizing and appropriating funds for each phase of each project as that phase is approved. Congress has already approved multiyear programs; but, at present, the limit provided by the Constitution for full funding is two years. Two-year authorizations and appropriations are not sufficient for most major program phases, as we have shown in Chapter 3, and the end of the two-year period may or may not come at the end or beginning of a program phase. A two-year fiscal period force-fits program approval into a two-year time cycle and does not do away with the traumatic funding discontinuity that usually occurs at the end of a fiscal period. Synchronizing funding approvals with program milestones will permit full-phase approval, full-phase management, and full-phase contracting for major systems. This assurance of funding availability, coupled with the use of detailed definition and design-to-cost engineering techniques, will permit fixed-cost, full-phase procurement of major systems.

To enhance life-cycle management, a long-term goal that the Congress should undertake is the formulation and eventual enactment of a *single, simplified, generic, durable statute that governs the procurement process*. Just as tax reform was needed, so procurement reform is needed. The law should not be considered without extensive analysis, simulation, testing, and public debate before enactment. It could include elements of deregulation to enhance true and free competition, remove socioeconomic objectives as a goal of government procurement, and embrace Presidential attention to and Congressional involvement in goal setting and long-range dynamic national planning for major systems. The law would consolidate and update past procurement laws, resolve conflicts between existing laws, and anticipate and embrace legislative needs for the procurement process for the 1990s and into the next century. The new law could provide a vehicle for restating public policy related to procurement of large systems, provide enforceable penalties for dishonest acts, and prepare for a new era in which the United States will make a *commitment to succeed*! The reason that many initiatives, remedies, and reforms of the past have not succeeded is that they have been done in haste to solve specific problems. They tried to cure the ills of major systems acquisition in a step function rather than with a gradual transition.

A Call for Legislative Restraint

Any suggested framework of positive actions to enhance efficiency in our major systems acquisition process must also include calls for restraint. Effective procurement reform must include avoiding importune changes in procurement policy or law. Those who are faced with complying with the many procurement laws already on the books are calling loudly for the policy makers and legislators to stop adding to the confusion by issuing more policies and enacting more legislation without thorough study of the full implication and effects of the new laws. An increasing number of procurement professionals in both industry and government are calling for legislative restraint. The energy saved through restraint can be redirected into a thorough study, analysis, and testing of new, durable, unified, simplified procedures and policies with their eventual adoption as they are proven workable and beneficial.

Experience over the past several years has proven that excellence in systems and projects cannot be legislated or directed, but that it must depend on sustained, proactive management throughout the organization doing the work. *Excellence can only be built into our systems and projects through sound management and engineering practices.* We must rely on our business and technical superiority rather than on legal remedies to achieve excellence.

The 1987 *Farmer's Almanac* contains the time-honored thought, "All progress stems from change but all change is not necessarily progress." We must avoid making change for its own sake, and we must revert to change only when it is needed to help accomplish long-term goals.

The 1971 final report of the Commission on Government Procurement, Major Systems Acquisition Study Group, contained a closing "prayer" as part of a section on national policy issues:

Closing Prayer

> *We believe that an entirely new set of buzz words should become the characteristic terminology for communication of intent in the system acquisition process.*
>
> *We suggest that these buzz words should include the terms,* people, competence, freedom, incentives, flexibility *and* trust. *This final recommendation is an effort to make the prior ones workable.*
>
> *The change that* must *occur if the other needed changes are to become effective, is an* attitude *change. Fairness, reasonableness and the judgment of competent people, together with recognition of the simple truth that authority and responsibility must go hand and hand, are essential to the realization of the improvement goals which this study can recommend but cannot put into effect.*
>
> *It must be acknowledged that these simple things have gradually disappeared from the systems acquisition process as they have tended to disappear from much of the Government Procurement scene. By expressing a need for their reinstatement we can view our suggestion as* radical *only in the sense that it would be such a tremendous change from the prevailing attitudinal environment.*

Why Moves to a New Environment Are Now Feasible

With all of the foregoing said, there are still the lingering questions of: Why haven't we been able to do anything so far in effectively executing the good ideas that have been proposed by generations of analysts and recommended in volumes of studies? And what makes us think we can succeed now?

The principal reason for inaction in acquisition reform has been lack of sustained public and Congressional support. Reforming any process or philosophy that has been growing for years is difficult. It takes both time and strong public support to unseat old ideas and to supplant them with new ones because of the inertia—and tradition—built up over the past years. Without the sustained interest and support of the public and of Congress, the recommended policy reforms will be difficult to achieve. In the past, the principal method of approaching reform in government procurement has been to establish a new commission, study group, committee, think-tank, or research contract to come up with new ideas and to propose new solutions. Once these solutions are developed, the group is disbanded, and the implementation of the recommendations is left up to a small staff working in the enormous bureaucracy. Outside groups, temporary commissions—both in the executive and legislative branches of government—and outside consultants have been brought in, instead of relying on the mainstream of people and organizations that must do the job on a day-to-day basis. Occasionally, reports and studies that have taken months or years to complete are discarded and almost forgotten. In other cases, the too-small staff left to implement the recommendations does not have sufficient clout to shepherd all of the needed reforms through the Congress and the implementing executive branch agencies. Then, when everyone realizes that the same problems are still with us, a new commission or committee of experts is formed to resurrect, combine, and provide new inputs to past studies. The process of forming new study entities and panels of experts every six or seven years provides the appearance of concern and action but often has too little long-term substantive impact on the acquisition process. It is for these reasons that we are convinced that (1) the need for reform must be recognized by citizens themselves and by their senators and representatives; (2) simplification and time-phased selective deregulation, new government-wide accounting and financial systems, tenured project management, and dynamic national long-term planning must themselves be methodically planned and presented to the people as an integrated approach; and (3) once approved by the electorate, these reforms must be planned, implemented, and operated by existing organizations in government rather than new ones.

There are a number of reasons why we can expect meaningful progress in these areas now even though past high-powered efforts have failed.

First, there is a heightened awareness, not only in this country but in others, of the detrimental economic effects of the continued huge federal deficit and the resulting steady growth of the national debt. Economists are informing the populace that personal economic well-being is dramatically improved by decreasing the amount of money owed by our country. Prospects of higher interest rates and tax increases are gaining taxpayer attention. Citizens, who are finding that government policies have a direct impact on their pocketbooks, are beginning to demand much more of their government in achieving efficiency, economy, and effectiveness.

Second, we now have better communication on public issues than we have ever had before. The almost insatiable appetite of the citizen for news is reflected in the greater attention to televised and broadcast Congressional hearings, cable news networks, news magazine shows on national as well as state and local issues, live satellite television broadcasts, national newspapers covering both business and general news, and even political satire and comedy. The news media provide speedy summarized information that is more easily comprehended and assimilated by the average citizen. Feedback television and radio shows with dial-in opinion polls can immediately assess public opinion and show the results in near real time. Communication technology, never before able to air public issues with such intensity, will bring forth issues such as the ones in this book with increasing regularity this year and next and on into the 1990s and beyond. Government will increasingly fall under the microscope of a demanding public and will be forced to respond rapidly and effectively.

Third, we are moving into a new generation of politically and economically savvy citizens. The "baby boomers" are now in their thirties and forties. Some are even looking forward to retirement and are concerned about the long-term economic well-being of our country and how it will affect their own retirement years and the lives of their children. Many are employed in the very industries that provide projects, products, and services that support our big federal systems and projects. They are increasingly well informed, well educated, and well read. This generation gives every indication of wanting to make lasting improvements in the society in which it exists, and it has the ability to do so.

Fourth, there are new breeds of high-technology, computerized, automated business systems becoming available almost every day. Computer hardware and software for accounting, scheduling, estimating, and cost control are now in their fifth or sixth generation. Massive supercomputers that operate thousands of times faster than conventional ones are now moving into the realm of supporting business and finance functions as well as engineering and scientific functions. Small desktop units now have the power of 1960s mainframe computers. Governments of some of the largest states have proven that accurate, credible cost accounting, tracking, and financial reporting are available to provide greater efficiency, economy, and effectiveness in their major projects and programs. Now, computer and accounting firms are beginning to study how similar systems can be upgraded to handle the mammoth financial tracking requirements of the federal government. The technology is now here to implement whatever policies, practices, and procedures are required, and to do it on a cost-effective basis. Rapid communication and data transfer, coupled with high-speed computer technology, has made feasible a government-wide computer network that can provide instant information on financial performance against budgets. This computer/communication system has not yet been developed because management philosophies and government policies have not yet been changed to make fruitful use of such a system. The technology exists, however, and is being further developed to respond to these needs as soon as meaningful reforms are adopted.

So there are some compelling reasons why significant moves could be made in the very near future which would result in dramatic progress toward simplification, efficiency, and economy in major federal acquisitions. Yes, reforms of this type are not easy, but many things are working together to produce an environment where change is not only possible but perhaps inevitable. It is up to each of us to continue to participate in and to support these efforts to bring about an orderly transition into a new era of efficiency, economy, and effectiveness in these systems and projects that are vital to our national well-being. Read the next section to see what you can do about it.

Endnotes

1. Comptroller General of U.S.: *Managing the Cost of Government*, General Accounting Office, Washington, DC, February 1985, AFMD-85-35.
2. Arthur Andersen: *Sound Financial Reporting in the U.S. Government*, Arthur Andersen & Co., Washington, DC, February 1986.
3. Arthur Andersen: *Guide for Studying and Evaluating Internal Controls in the Federal Government*, Arthur Andersen & Co., Washington, DC, August 1986.
4. Price Waterhouse: *Enhancing Government Accountability*, Price Waterhouse, New York, 1983.
5. Stephen Hess: *Organizing the Presidency*, Brookings Institution, Washington, DC, 1976.
6. Richard P. Nathan: *The Plot that Failed: Nixon and the Administrative Presidency*, Wiley, New York, 1975.
7. Richard Rose: *Managing Presidential Objectives*, Macmillan, New York, 1976.
8. General Accounting Office: *Joint Major System Acquisition by the Military Services: An Elusive Strategy*, General Accounting Office, Washington, DC, December 1983, NSIAD-84-22.
9. General Accounting Office: *Department of Defense Acquisition Case Study of MIL-STAR Satellite Communications System*, General Accounting Office, Washington, DC, 1986, NSIAD-86-45S-15.

10. Rodney D. Stewart: *Major Systems Planning and Budgeting in Support of National Goals*, George C. Marshall Space Flight Center, National Aeronautics and Space Administration, Huntsville, AL, 1972.
11. Presidential Commission on Space Shuttle: *Report of the Presidential Commission on Space Shuttle Challenger Accident* Presidential Commission on Space Shuttle, Washington, DC, June 1986.
12. Benjamin S. Blanchard and Wolter J. Fabrycky: *Systems Engineering and Analysis*, Prentice-Hall, Englewood Cliffs, NJ, 1981.
13. General Accounting Office: *Budget Issues Cost Escalation on Three Major Department of Transportation Projects*, General Accounting Office, Washington, DC, July 1986, AFMD-86-31.
14. Rodney D. Stewart and Richard M. Wyskida (Editors): *Cost Estimator's Reference Manual*, Wiley, New York, 1987.

Final Words: What *You* Can Do About It!

If you believe that fundamental changes need to be made in management philosophies, laws, rules, regulations, or incentives in the acquisition of major systems and projects, *you* can do something about it by communicating directly with members of Congress. This communication can take the form of letters, mailgrams, telegrams, telephone calls, or face-to-face meetings. The U.S. Chamber of Congress published the following list of "Dos" and "Don'ts" on communicating with members of Congress[a]:

Do

- **Do** identify clearly the subject or subjects in which you are interested, not just House and Senate bill numbers. Remember, it is easy to get a bill number incorrect.
- **Do** state why you are concerned about an issue or issues. Your own personal experience is excellent supporting evidence. Explain how you think an issue will affect your business, profession, community, or family.
- **Do** restrict yourself to one or at most two topics. Concentrate your arguments.
- **Do** put your thoughts in your own words. This is especially important if you are responding to something you read. If a member of Congress receives numerous letters with nearly identical wording, he or she may discount them as part of an organized pressure campaign. Even so, pressure campaigns have worked when mail was so voluminous that it had to be weighed rather than read!
- **Do** try to establish a relationship with your own representative and senators. In general, you'll have more influence as a constituent. If you

[a] Reprinted with the permission of the Chamber of Commerce of the United States of America from *A Guide to Communicating with Members of Congress*.

don't know whose district you are in, call your local county voter registrar and give him or her your zip code.

- **Do** communicate while legislation is being considered by congressional committees and subcommittees, as well as when it is on the House and Senate floor.
- **Do** find out the committees and subcommittees on which your representative and senators serve. Members of Congress have much more influence over legislation within their committees' and subcommittees' jurisdiction.

DON'T

- **Don't** ever, ever threaten. Don't even hint "I'll never vote for you unless you do what I want." Present the best arguments in favor of your position and ask for their consideration. You needn't remind a member of Congress of electoral consequences. Mail and phone calls will be counted without your prompting.
- **Don't** pretend to wield vast political influence. Write member(s) as a constituent, not as a self-appointed spokesman for your neighborhood, community, or industry. However, if you really are a spokesman for a group, be sure to mention it.
- **Don't** use trite phrases or cliches. They can make your letter sound mass produced when it isn't.
- **Don't** become a pen pal. Some congressional offices don't bother to count mail from seemingly tireless letterwriting constituents.

I Suggest the Following Subjects for
Congressional Attention:

1. Dynamic, Long-Range National Planning ☐
2. A Uniform Government-Wide Accounting System Tied to the National Budget ☐
3. Increased Tenure for Project Managers ☐
4. Simplification of the Procurement Process ☐
5. Pre-Legislative Analysis, Simulation, and Testing Before Enactment of Procurement Laws ☐

Signed ______________________________

Where to Get Additional Information

American Enterprise Institute for
Public Policy Research
1150 17th Street NW
Washington, DC 20036
(202) 862-5800

American Society of Macro
Engineering
Polytechnic Institute of Brooklyn
333 Jay Street
Brooklyn, NY 11201
(718) 643-7170

American Society for Public
Administration
1120 G Street NW Suite 500
Washington, DC 20005
(202) 393-7878

Association of Government
Accountants
727 South 23rd Street
Arlington, VA 22202
(703) 684-6931

The Brookings Institute
1775 Massachusetts Avenue NW
Washington, DC 20036
(202) 797-6000

Center for Defense Information
1500 Massachusetts Avenue NW
Washington, DC 20005
(202) 862-0700

Center for National Security
Studies
122 Maryland Avenue NE
Washington, DC 20002
(202) 544-5380

Coalition for Common Sense in
Government Procurement
1990 M Street NW Suite 400
Washington, DC 20036
(202) 331-0975

Common Cause
2030 M Street NW
Washington, DC
(202) 833-1200

Department of the Treasury
15th & Pennsylvania Avenue NW
Washington, DC 20220
(202) 566-5252

Ethics Resource Center
1025 Connecticut Avenue NW
Washington, DC 20036
(202) 223-3411

Federal Acquisition Institute
18th & F Streets NW
Washington, DC 20405
(202) 535-7788

Fund for Constitutional
Government
P.O. Box 15007B
Washington, DC 20003-0007

General Accounting Office
P.O. Box 6015
Gaithersburg, MD 20877
(202) 275-6241

House Armed Services Committee
2120 Rayburn Office Building
Washington, DC 20515
(202) 225-2191

Institute of Cost Analysis
7111 Marlan Drive Suite A
Alexandria, VA 22307
(703) 768-6405

Institute for Policy Studies
1901 Q Street NW
Washington, DC 20009
(202) 234-9382

Joint Financial Management
Improvement Program
666 11th Street NW Suite 705
Washington, DC 20001
(202) 376-5415

Large Scale Programs Institute
2815 San Gabriel
Austin, TX 78705
(512) 478-4081

Military Operations Research
Society
101 S. Whiting Street Suite 202
Alexandria, VA 22304
(703) 751-7290

National Committee for an
Effective Congress
400 C Street NW
Washington, DC 20002
(202) 547-1151

National Contract Management
Association
6728 Old McLean Village Drive
McLean, VA 22101
(703) 442-0137

National Estimating Society
1001 Connecticut Avenue NW
Suite 800
Washington, DC 20036
(202) 466-5499

National Planning Association
1616 D Street NW Suite 400
Washington, DC 20036
(202) 265-7685

National Taxpayers Union
325 Pennsylvania Avenue SE
Washington, DC 20003
(202) 543-1300

Procurement Round Table
1001 N. Highland Street Suite 201
Arlington, VA 22201
(202) 393-1780

Project Management Institute
P.O. Box 43
Drexel Hill, PA 19026
(215) 622-1796

Project on Military Procurement
422 C Street NE
Washington, DC 20002
(202) 543-0883

The Rand Corporation
P.O. Box 2138
Santa Monica, CA 90406-2138
(202) 296-5000

TAG: The Association Group of
Small Research, Engineering,
and Technology Companies
P.O. Box 9411
McLean, VA 22102
(301) 564-8873

U.S. Chamber of Commerce
1615 H Street NW
Washington, DC 20062
(202) 659-6000

APPENDIX A

An Interview with Dr. Eberhard Rees

RDS: *The first question we'd like to ask you is, "If you could think back from this vantage point, what do you think would be the most important lesson that you learned throughout the space program and through your experience in managing and developing high technology systems?"*

DR. REES: *This could be pages! Lessons that I have learned especially? You ought to listen to everybody first! And not say, "Ah, well, this is not important." Just listen to everybody! And then come to certain conclusions from what you have heard, especially in talking to contractors. And I have some experience there. You must remember that, number one, the contractor is there to make money. There's no doubt of it, and you should be always aware of this when you are with the government as I was: that their motive is, number one, to make money. And there's nothing wrong with it; otherwise, the contractor would not exist. Ok! So they try everything to make money. On the other hand, there are many schemes to make money which might hurt the project or might hurt the taxpayer or might hurt the government. So this is what you have to find out: What principles should be observed when you make a contract, and how you should supervise them. The next point is that we, from the government side, having had this contract (not only prime contracts but subcontracts also) with the aerospace industry had always felt we should have our engineer located on-site with the contractor. This was, in my opinion, MOST IMPORTANT.*

RDS: *To have someone on-site to really penetrate . . .*

DR. REES: *Yes, someone on-site. It turned out that these people (engineers) on-site very often sided with the contractor. The contractors did not like co-located engineers in the beginning at all. I remember particularly that some Boeing key personnel were strongly against it. Also other people from McDonnell Douglas on the Skylab were strongly against it. But later, they found out that these co-located engineers are very helpful, that they wouldn't cost them any money, and would help them greatly. On the Skylab especially we had a very very close relationship with McDonnell Douglas which was very very helpful. So we have found out that this is MOST IMPORTANT—getting our good engineers in there with the contractor and working together with them as if they were contractor personnel. Just thinking of the project and not 'Am I a civil service man or am I a contractor's man' . . . but just what is good for the project?*

ALS: *In your time at NASA, what projects did you think were most successful from a procurement standpoint and why?*

DR. REES: *Overall the most successful program we ever had was, of course, the Apollo program. There is no doubt about it! It was a very difficult thing to do because we had to make this big, big, big rocket that would go to the moon. Of course going to the moon wasn't easy and it was a complicated thing. The only difference, compared to the shuttle, was this: that at every point on the whole trajectory and while on the moon and back from the moon they were always able to escape. On one mission, if you might remember that, we had a failure of some maneuvering rocket . . .*

RDS: *That was during the flight or on orbit. . . .*

DR. REES: *Yes! And they just came back! Only we didn't land on the moon as planned. This is why we had only six moon landings instead of seven (in the Apollo program).*

RDS: *And I think the Apollo was very successful from a cost standpoint too, wasn't it?*

DR. REES: *Yes!*

RDS: *Somebody estimated $20 billion dollars at the start?*

DR. REES: *You know President Kennedy at that time just said we are going to go to the moon; and the Congress and the Senate said yes! There was no talk about money. Finally Kennedy came to [Dr.] Seamans and said, "Well, how much would it cost?" And then Seamans said, "I'm sorry but we do not even know how to go to the moon yet. So it will be between $20 and $40 billion." And Kennedy said, "All right, fine." And nobody—this would not happen today—nobody ever questioned it! Then, finally, the only lady later on who questioned it was the senator from Maine—what was her name?*

ALS: *Oh, yes, Margaret . . .*

DR. REES: *Margaret Chase Smith. And she was quite sharp and wanted to keep us on $18 billion because this was a five-year estimate we made sometime before. The Apollo program cost about $23 billion finally.*

RDS: *That was very good estimating!*

DR. REES: *That was very good. So from a cost point of view (the Apollo program) was not too bad.*

RDS: *OK That's great. Now the other part of our question is..*

ALS: *The biggest failure..*

RDS: *The least successful program in terms . . . I guess we're talking about not just technically but maybe cost-wise, overruns, things like that?*

DR. REES: *Of course as long as I have been here in this country we had (mostly successes). In El Paso, Texas, we prepared written reports for a long time and tried out the V-2 at White Sands Proving Ground. We made high altitude shots for the scientists there. And the first V-2 we shot into Mexico. People said, "These Germans, they have done it on purpose." (laughter)*

RDS: *Oh, they thought you had done it on purpose?*

DR. REES: *Yes! And then everybody was running over to Mexico to get the debris and they had* sold *them already! (laughter)We didn't get too much of that rocket back. I don't remember exactly what went wrong there. It was in the guidance system. Of course, we didn't hit a city or didn't kill anybody.*

Then the Korean War started, and it was said we ought to go deeper into the development of guided missiles. We started a project of making a bigger guided missile out of the V-2. Texas was too small so we had to go to Alabama. (laughter) At that time there wasn't any possibility in Fort Bliss, Texas, to expand, and Redstone Arsenal was free at that time. I remember in July 1949 we made a trip from there to here and looked the place over and said, "Well, this is fine." It is on the Tennessee River so we could ship bigger missiles from here down to Florida. So we felt this was a good thing: to go to Huntsville. A lot of people from Huntsville did not feel that way at that time because, well, it didn't fit [into the lifestyle of] these cotton people here. This was something entirely different than they were used to. We came here when the Army was building up at Redstone Arsenal. Thiokol was already here and Romm and Haas. I remember [Dr.] Hank Shuey was already out running around and he was building a house up here [on Monte Sano] where he still lives [Dr. Shuey died in 1987]. So I got acquainted with him very early. Then we made the Redstone missile. And the Redstone Missile was, I think, even fielded. . . .

RDS: *That was a pretty successful project—the Redstone?*

DR. REES: *The Redstone was a successful project. I do not know about the money that it cost. . .*

RDS: *We're planning some research to see if we can find out! [We later discovered that the budgeted cost was $76.5 million. Actual costs were not obtainable because the funds came from so many sources.]*

DR. REES: *So after the Redstone missile came the Jupiter missile! The Jupiter missile was in competition with the Air Force Thor. Of course, if we were to develop a much bigger missile, then we had to have bigger facilities. We had built one test stand already for the Redstone Missile. At that time (General) Toftoy was in charge, and he had agreed right away to make that test stand much bigger for much bigger missiles. So that first big test stand was going to be all right for the Jupiter. And then we went into the work of the Jupiter. We always felt the Jupiter would have been a better missile than the Thor project but the Army lost against the Air Force. That was General Medaris against General Shriver. And Shriver just won.*[a]

We even launched a shot to the moon with the Jupiter! People have forgotten it entirely. I brought it up only in a speech I had to make on [Dr.] Lucas' (retirement) celebration. I brought it up since nobody in the room remembered that we shot to the

[a] See Michael H. Armacost: *The Politics of Weapons Innovation: The Thor-Jupiter Controversy*, Columbia University Press, New York, 1969.

moon very early. That was in 1959 under the Army when we did not have Marshall Space Flight Center.

RDS: *Was that a circumlunar flight?*

DR. REES: *No, it was just going past the moon. We missed by 30,000 miles, I remember. Then, you know, [Dr.] Von Braun was always thinking of space flight all the time. And Toftoy, he couldn't do too much about it actually. It would have been misappropriation of funds. So very few people (supported the space flight idea): Kraft Ehrike left us finally because he was just a space flight guy and because he couldn't do too much in space. And Medaris, when he came here was more open [to space flight possibilities], and when Von Braun suggested that the Redstone Missile be equipped with a payload—not a big one—for space flight, we went rather far with JPL [Jet Propulsion Laboratory]. The JPL people were also very strongly for space flight all the while. There was Dr. Froehlich coming over here and Dr. Pickering and all those people. So we went rather far in that whole thing.*

[Dr. Rees explained here, in a nonrecorded portion, how the team got involved in space flight after the Russian Sputnik launch.]

DR. REES: *At that time we detected, with Explorer I, the Van Allen Belt. You know that payload of thirty-two pounds, we made together with (Dr.) Van Allen. And it was the Van Allen belt we detected: so this was a much greater feat than the Russians, who had only their payload up there with no scientific instrumentation.*

RDS: *It didn't really do anything did it? It just was a big payload.*

DR. REES: *Yes! And then when we launched that Jupiter later on toward the moon, we detected the second Van Allen belt which was farther out there. So Explorer I was a very successful one and didn't cost too much money. We should not forget that the Army made space flight in the United States first because it was the Army, not the Air Force, that had about four or five of these Explorers up there already. I remember we launched a nuclear payload from Johnson Island which is about 800-1000 miles west of Hawaii. We launched that with a Jupiter missile, and parallel to that we launched an Explorer to find out what radiation that shot would cause. So, after my opinion, under (General) Medaris we did a lot of interesting scientific things. Of course Medaris wanted to stay in space but the Army just didn't feel they should stay in space. Only Secretary Brucker, Secretary of the Army, was for it; but his generals, some of them, were not too strong about it. And then the Air Force. . . .*

RDS: *The Air Force could see that this would fit into their mission better?*

DR. REES: *Yes! Jupiter was then cancelled. We launched some satellites with Jupiter II with much higher payloads of course. And . . . the Thor missile remained with which the Air Force put up some satellites. But all these projects under Medaris were, in my opinion, very efficient from a money point of view and from the technical side. And then in July 1960 came the Mercury programs where we put Shepherd up—not into space but just man sitting on a rocket. It was also inexpensive and was successful.*

RDS: *We had a suborbital flight first, I believe, and then we had an orbital flight.*

DR. REES: *Yes, before that we launched a monkey. And it was that monkey that died about a year ago. . .*

ALS: *Miss Baker. . . .*

DR. REES: *It is just too bad that we have not made any point of that. This type of squirrel monkey lives about eight years and that one lived, I think, almost twenty years: so you could have said, "Well, that was because of space."*

RDS: *Space flight lengthens your life? (laughter)*

DR. REES: *It was only a theory. So then came the Mercury program and the Gemini program. In that program [Gemini] we were not too deeply involved.*

RDS: *That went up on an Atlas, didn't it?*

DR. REES: *Yes, that was on an Atlas. And then came the Apollo program. As to the Apollo program, we can only say this was successful so there was, so far, no bad program. So the answer so far is no! After that came the Skylab and that worked all right too. That was very close to failure too you know, some of those solar cells broke off on one side. That was quite dangerous but it worked. Of course under Rocco Petrone we were working well and we had a real space station up there. So we, in a sense, had a space station already. And if they would have put a little bit more money in that thing [Skylab] we still would have that space station up there. I remember exactly that for a cost of about $300 more million we would have launched the second one which is now in Washington in the museum, but we would have had to put on more expendables and more boosters simply because you know they come down again sometimes unless you boost them up. But that was a space station and it was a two-room craft occupied by the astronauts, three times bigger than the Russian Salyut. We launched it up with the Saturn V, and out of the S-IV stage, we made that spacecraft. This was a good project and we had a space station up there, no doubt!*

RDS: *That included the "workshop" and then they had the ATM along with it: the Apollo Telescope Mount. . . .*

DR. REES: *Yes! So this was, again, a good project but then just $300 more million were turned down because Washington at that time said, "We'd rather put that money into the shuttle." Because we estimated at that time that the shuttle would cost about $8 billion for the development and I remember (President) Nixon had only $5.3 billion. So we really didn't make [the space shuttle] as big and as complicated as we had the intention to do at first, and out of that came the next decision, to put on these two solid boosters. It was decided at that time (Dale Myers was the man in charge . . . George Miller and Phillips left after the Apollo program. . . then we had Dale Myers there [for manned space flight] and it [NASA] was under Fletcher at that time) to put these two solids on. This [arrangement] was used already with the Titans. At that time there was much talk about "Should we make these in one piece" and so forth. But it was said, "This just can't be handled." So we made that separation. Unfortunately it led to the [shuttle] accident.*

[Here, in an unrecorded portion of the interview, Dr. Rees discussed the Shuttle accident and its causes.]

DR. REES: *I was in Washington four weeks ago and Fletcher called the "old guys" together. There was (General) Sam Phillips, George Miller. There was even Brainard Holmes and Bob Gilruth. It was a very interesting meeting and we talked about it. It was just the day before the Titan Air Force rocket exploded [April 18, 1986]. We were just sitting there and said, "Now we don't have any rocket at the present time to put important payloads up there." And then I made the suggestion, "Let's launch another shuttle as it is, because if we would observe the temperatures and really check it out properly those two o-rings which worked 24 times, in an emergency, could work again . . . we could launch another four or five if we really know what happened and then go into that." But everybody said this was politically entirely impossible after the [Presidential] Commission[1] was set up.*

DR. REES: *And so we are now on the shuttle. Going all the way from the beginning and so far the answer is no. [We had no really unsuccessful programs].[b] And [in spite of] that shuttle accident. . . in my opinion the shuttle was also a good accomplishment. Just that particular part was overlooked and Marshall (Space Flight Center) had been accused very strongly. The only accusation you can really make is that it seems that a lot of people at Marshall had known this as a weakness for some time and that really something should have been done years ago.*

RDS: *I think it goes back to maybe the first thing you said that having competent engineers* throughout *the project not only at the government installation but also at the contractors. . . .*

DR. REES: *You know we had, at that time (during the Apollo Program), 7,300 people at MSFC. And then Von Braun had to lay off 750 and then it was my turn and I had to lay off about 1,200 and then Petrone and Lucas [subsequent MSFC Directors] had to lay off so many that it [staffing] is now, at the present time, 3,100. And this is smaller than 50 percent. So they have only about 40 percent of the manpower. . . . and of course when you lay off people you lay off a lot of very outstanding people. Ludie Richards, for instance, and people like this just left. So it is not only the quantity [of people that is important] but also the quality, there's no doubt about it!*

Endnotes

1. Presidential Commission on the Space Shuttle: *Report of the Presidential Commission on Space Shuttle Challenger Accident*, Presidential Commission on Space Shuttle, Washington, DC June 1986.

[b] Dr. Rees is speaking of all of the missile programs and rocket propulsion programs from the time the German scientists came to the United States until his retirement January 1973.

APPENDIX B

The Organization and the Role of the Project Manager

The program manager and the importance of the function he performs has received a growing recognition and acceptance in the acquisition of major systems. This has been the case *for many years* in DOD weapon systems due to the ever-increasing sophistication, complexity, urgency and national significance associated with most major defense programs. Other agencies, such as the AEC, NASA, and DOT have followed the DOD approach on their more significant acquisitions and the currently emerging civil agencies are also recognizing that the individual program management office provides advantages over the straight functional organization. In fact, many civil agencies have wisely drawn upon personnel with expertise and experience gained while managing major DOD programs.

Although the role of the program manager has gained wide acceptance and recognition, it is quite evident that many of the management responsibilities upon which program success depends is guided and influenced at many levels of authority above the program manager. Such guidance and influence *starts with the Congress* and extends down through the Executive with many layers of staffs and management involved. In recognition of this situation, the Task Team chose first to make recommendations concerning the Role of Congress in the Management of Major Systems Acquisition, The Delineation of Principles and Policies by the Executive Branch, and The Key Decision and Review Process. Unless action is taken to clarify responsibilities at all levels, delineate broad principles and policies, and further enhance the role of the program manager as recommended in this report, programs will continue to experience the problems so predominant in the past.

Note: Appendix B is an excerpt from the final report of the Major Systems Acquisition Study Group of the Commission on Government Procurement, January 1972. (Emphases added throughout.)

Assuming that action will be taken on recommendations involving management at levels above the program manager, the Task Team devotes this section of its report on steps which should be taken to improve the management of major acquisitions at the program office level. Although the agency organization and the role of the program manager varies greatly between agencies, within agencies and even with respect to different phases of a single project, it is universally accepted that program success is impossible without providing appropriate authority and organizational support to the program manager so that good day-to-day decisions are possible at the "get the job done" level. It is also recognized that authority must provide the flexibility necessary for essential trade-offs and that direct communications and control are paramount. Our studies concentrated upon a detailed analysis of these considerations.

The findings, conclusions and the recommendation of the Task Team concerning the Organization and the Role of the Program Manager are stated below.

1. *Findings.*
   a. Program success is frequently jeopardized through lack of a definite charter for program management.
   b. Program managers are often put into an untenable position because of decisions made early in the acquisition process and prior to their individual selection and organization of their program office. Such decisions usually occur in the requirements determination, acquisition planning and systems conceptual phases.
   c. The program managers often do not have complete responsibility and authority for all decisions at the program office level.
   d. The program manager does not necessarily have to be his own contracting authority to effectively implement his decisions.
   e. The program manager often does not have the same degree of control in contracting that he has in other areas of responsibility. Proper safeguards are necessary through legal reviews, procurement committees, and contract approvals reserved at higher levels to ensure statutory and regulatory compliance.
   f. Both the matrix and vertical organizational approaches to major systems acquisition have proven to be successful.
   g. The matrix organizational approach is the more economical and leads to cross feed of technology and experience between programs.
   h. The availability of sufficiently qualified personnel is a problem in major systems acquisition management and dictates the maximum use of the matrix approach, with the vertical approach reserved for use in unusual cases.

i. Civil service regulations do not provide adequate flexibility to allow program management to make maximum use of available civilian talent.

j. There is a need for career fields for those involved in program management, improved training programs, and better personnel motivation. The policy of periodic reassignments often acts to the detriment of programs.

k. Technical support, including system integration, obtained through the use of government capabilities, not-for-profits, and prime/principal contractors have all proven to be successful approaches to program management.

l. Government and industry officials both endorse the maximum use of government in-house capability for technical support in program management. When the government's capability is not adequate, then the prime/principal contractor capability is preferable to the use of Federal Contract Research Centers or not-for-profit organizations.

m. Third party reviews have proven to be beneficial on specific identifiable problems when knowledge of the total acquisition is not necessary and when the decision making is left to the program manager.

n. Reviews by higher authority regarding problems which should be resolved at the program manager level tend to delay decision making.

2. *Conclusions.*

a. Program management has been most effective when there is a definite charter and when the manager has been guided by broad principles. The issuance of the charter eliminates uncertainty as to scope of program responsibility and lines of authority.

b. Program managers should be assigned to a program early in the acquisition process and prudent forethought should be taken in the making of decisions early in the program, especially those prior to full scale development.

c. The program manager must be delegated authority commensurate with his responsibilities for the management of his program and the key decision process should provide sufficient flexibility to insure continuing trade-offs for performance, schedule and cost at his level.

d. The matrix approach has provided the more economical organization for the procuring agencies in major systems acquisition when human resources are limited.

e. Technical support, including technical integration, through the use of internal capabilities, not-for-profit organizations and prime/

principal contractors have all proved to be successful, but the use of in-house capabilities is most suitable. When in-house capabilities are not sufficient, the prime/principal contractors or not-for-profit organizations should be used in that respective order.

f. The program manager need not necessarily be delegated contracting authority to effectively implement his decisions; however, he should have control over the procurement activities which relate to his program.

g. Third party reviews can be utilized effectively by program managers on specific identifiable problems where knowledge of the total acquisition is not necessary.

h. Program managers success is dependent upon the competence and motivation of the personnel assigned to the program. Career status, additional training, and motivation of personnel are required to improve the effectiveness in the management of major systems acquisition.

i. Civil service regulations do not recognize program management as a career field nor provide sufficient incentives and flexibility for the efficient use of available civilian personnel.

3. *Recommendation.*

a. A firm and separate charter for program management be issued on each major systems acquisition.

b. Appoint program managers early in the acquisition process and allow them to participate in decisions made during the requirements determination, acquisition planning and system conceptual phases.

c. Maximum use be made of the strong matrix organization by the agencies as the approach to major systems acquisition and that the use of the vertical organization be reserved for those acquisitions of great national importance which involve an unusually high degree of uncertainty/risk.

d. Program managers be given direct responsibility, authority, and in-house support essential for making timely and continuing trade-off decisions between cost, schedule, and technical considerations.

e. Maximum use be made of in-house capability for technical support, including technical integration. If necessary, technical support be contracted with a prime or principal contractor when the contractor possesses the capability and that Federal Contract Research Centers or not-for-profit organizations be used only on a case-by-case basis.

f. The program manager be given control in the contracting area similar to that which he is given in other functional areas.

g. The use of third party reviews be utilized only on a select basis when identifiable problems exist.
h. All agencies establish career fields for those individuals involved in program management.
i. Better training programs, longevity on the job and employee motivation for both military and civilian personnel be provided and that steps be taken to provide the flexibility essential for managers to make maximum use of available talent.

APPENDIX C

The Role of Congress in the Management of Major Systems Acquisition

The Congress is by far the most significant potential force to improve the management of major systems acquisition. This belief is based on the material we have studied, presentations we have listened to, and comments received from our in-depth interviews with a wide spectrum of ranking Government and industry officials. Of the seventeen topics considered in the interviews and mailed to the participants in advance, only one dealt specifically with the role of Congress. However, everyone with whom we talked was interested in the role of Congress and many felt quite strongly concerning the subject.

Since those who were interviewed are leaders in the field of major systems acquisition, all of them recognized the responsibilities of Congress and no one questioned its authority. They did believe certain changes could be made by Congress which would greatly improve the procurement process for major systems. At the end of each interview, we usually asked the question, "Based on your knowledge and experience, what one recommendation would you make that in your opinion would accomplish the most to improve the management of major systems acquisition?" The majority of replies dealt with recommendations to eliminate uncertainties and instabilities in authorization/funding and *to clarify the decisionmaking and legislative process* involved in overall management.

The findings, conclusions and the recommendation of the Task Team concerning the Role of Congress in the Management of Major Systems Acquisition are stated below.

1. *Findings.*
   a. There is total support in both the Executive Branch and industry regarding the fact that the role of Congress must be all-encompassing if it is to fulfill its responsibility to the public.

Note: Part of Appendix C is an excerpt from the Final Report of the Major Systems Acquisition Study Group of the Commission on Government Procurement, January 1972. (Emphases added throughout.)

b. Congress, through its actions or inactions has a profound effect upon the management of major systems acquisition.
c. The role actually reserved by the Congress in major systems acquisition decision making is not clearly delineated and fluctuates from time to time and on a case-by-case basis.
d. There is confusion with respect to the roles of the various Congressional committees and the related workings of the legislative process.
e. Committee staffs often lack sufficient expertise and manpower to cope with the volume and complexity of major systems being considered.
f. Congress does not always clearly state its requirements for program information and then depend upon the Executive Branch to fulfill its requirements.
g. There is an increasingly greater Congressional dependence upon the General Accounting Office which is, therefore, expanding its technical and systems analysis capability so that it can participate in the program decision process. This greatly expands its customary role of the past.
h. Uncertainty and instability of program authorization and funding along with untimely legislation has adverse effects on the management of major systems acquisition.
i. The dual system between the Executive Branch budgeting process and the Congressional authorization/appropriation process does not always lead to clear decision making between the Executive Branch and the Legislative Branch.
j. Although some flexibility results from a dual budgeting system, there is a need for Congress to formally recognize the necessity for flexibility within the Executive Branch to cope with such factors as program uncertainties, systems technical growth and other future events inherent in major systems acquisition.
k. Long range planning, e.g., Five Year Defense Plan, does not receive adequate consideration and endorsement by the Legislative Branch in authorization and appropriation processes.
l. *Cost growth on major systems acquisition can be caused as much by Congress in its role as by actions with the procuring agency*
m. Life-cycle costing early in the major systems acquisition process without due consideration of uncertainties, inflation and unknowns often inhibits program success.
n. Adversary relationships among Congress, the Executive Branch and industry often result in management decisions which are not in the best interest of either the program involved or the Nation.

2. *Conclusions.*
   a. The role of Congress in major systems acquisition is all-encompassing and must continue to be so if the Legislative Branch is to fulfill its responsibility to the public. *Members of Congress have become increasingly involved in the details of managing systems programs; hence, they should recognize and appropriately promulgate the important, special nature and unique management considerations associated with major systems acquisition.*
   b. The role of Congress in the management of major system programs needs to be clarified and thoroughly communicated. Although the interest of Congress is a shifting thing, responsive to wide swings in public opinion and subject to outside influence, much can be done to provide a better working relationship between the Legislative and Executive Branches.
   c. The Congress has become increasingly involved in decisions which require significant amounts of detailed technical information of a system analysis nature for sound judgment and decisions. The General Accounting Office is not presently capable of providing such data, nor should such a capability necessarily be developed in the GAO or in a new agency of the Congress when it is already available within the Executive Branch.
   d. The uncertainty, instability and the untimeliness associated with program authorization and funding are major causes of many problems being experienced in the management of major systems acquisition.
   e. The need for flexibility in the application of appropriated monies should be formally recognized by Congress so that the Executive Branch is better able to cope on a continuing basis with changes and reprogramming associated with program uncertainties, system technical growth and other future events inherent in major systems acquisition.
3. *Recommendation.*

   It is recommended that the role of Congress in the management of major systems acquisition be clarified and enhanced by accomplishing the following:
   a. The importance, special nature and unique management considerations of major systems acquisition be recognized, established and appropriately promulgated by the Congress and that a realistic approach to major systems acquisition be reflected in related legislation.
   b. The role of Congress and its committees in the management of major systems acquisition be more clearly delineated.

c. Congressional needs for program management information be fulfilled by the appropriate agency of the Executive Branch instead of developing a separate technical and systems analysis capability within the Legislative Branch.

d. Authorization and appropriation legislation be more timely, on a multi-year basis, provide more flexibility in the application of appropriated monies to various programs and contain fewer restrictions regarding the program management process.

Bibliography

Acker, David D.: "Reform 88": A Program to Improve Government Operations, *Program Manager*, U.S. Government Printing Office, Washington, DC, March–April 1983.

Adams, John R., C. Richard Bilbro, and Timothy C. Stockert: *An Organization Development Approach to Project Management*, Project Management Institute, Drexel Hill, PA, 1986.

Aerospace Corporation: *System Segment Specification for Communications Segment of CSOC*, Aerospace Corporation, Los Angeles, CA, December 1983, CSOC-00004.

Aerospace Industries Association: *Aerospace*, Aerospace Industries Association, Washington, DC, Spring 1984.

AIAA: "Aerospace America," *Aerospace America*, AIAA, New York, September 1986, Vol. 24, No. 9.

Aines, Andrew A.: *Federal Government Does Deserve Excellence*, Government Computer News, Government Computer News, Silver Spring, MD, November, 21, 1986.

Alexander, Tom: "Making High Tech Weapons for Less," *Fortune*, February 1985.

American Metal Market: "GAO Savings Disappointing for Multiyear Procurements," *American Metal Market*, Fairchild Publishers, Inc., New York, June 1983, Vol. 91, No. 13.

American Metal Market: "Alter Funding Formula to Multiyear Funding," *American Metal Market*, Fairchild Publishers, Inc., New York, March 1984, Vol. 92, No. 19.

Analytic Sciences Corporation: *AFSC Cost Estimating Handbook Series*, Analytic Sciences Corporation, Reading, MA, 1985.

Andersen, Arthur: *Sound Financial Reporting in the US Government*, Arthur Andersen & Co., Washington, DC, February 1986.

Andersen, Arthur: *Guide for Studying and Evaluating Internal Controls in Federal Government*, Arthur Andersen & Co., Washington, DC, August 1986.

Anderson, James H.: "Peacekeeper Basing Team," *Military Engineer*, July 1986, No. 507.

Anderson, Lane K.: *Accounting for Government Contracts: Cost Accounting Standards*, M. Bender, New York, 1981.

Anderson, Lane K.: *Accounting for Government Contracts: Federal Acquisition Regulation*, M. Bender, New York, 1985.

Anthony, Robert N., & Ross Graham Walker: "Cost Allocation," *Journal of Cost Analysis*, Institute of Cost Analysis, Washington, DC, 1984, Vol. 1, No. 1.

Aquidneck Data Corporation: *Aquidneck Data Corporation: A Working Philosophy*, Aquidneck Data Corporation, Middletown, RI, 1986.

Arnavas, Donald P., and William J. Ruberry: *Government Contract Guidebook*, Federal Publications, Washington, DC, 1986.

Association of Government Accountants: *Strengthening Controllership in the Federal Government*, Association of Government Accountants, Washington, DC, May 1985.

Auer, Joseph, and Charles E. Harris: *Major Equipment Procurement*, Van Nostrand-Reinhold, New York, 1983.

Augustine, Norman R.: *Augustine's Laws, 3rd ed.*, Viking, New York, 1986.

Augustine, Norman R.: "Inexorable Law of the Bungle," *New York Times*, New York Times, New York, March 1986, Vol. 135, No. 16.

"Aviation Week and Space Technology," *Aviation Week*, McGraw-Hill, New York, November 3, 1986.

Baldwin, William L.: *The Structure of the Defense Market, 1955–1965*, Duke University Press, Durham, NC, 1967.

Bateman, Lynn: "Factors Conspire Against Shorter Procurement Cycle," *Government Computer News*, Government Computer News, Silver Spring, MD, September 26, 1986.

Baumgartner, J. Stanley (Editor): *Systems Management*, Bureau of National Affairs, Washington, DC, 1979.

Bedingfield, James P., and Louis I. Rosen: *Government Contract Accounting, 2nd ed.*, Federal Publications, Washington, DC, 1985.

Biery, Frederick P.: "Cost Growth and the Use of Competitive Acquisition Strategies," *National Estimator*, National Estimating Society, Washington, DC, Fall 1985.

Biery, Frederick P.: "Accuracy of Military Cost and Schedule Forecasts," *Journal of Cost Analysis*, Institute of Cost Analysis, Washington, DC, Spring 1986, Vol. 3.

Black, Norman: "$2.2 Billion Contract Nears OK," *Huntsville Times*, Huntsville Times, Huntsville, AL, September 26, 1986.

Blanchard, Benjamin S., and Wolter J. Fabrycky: *Systems Engineering and Analysis*, Prentice-Hall, Englewood Cliffs, NJ, 1981.

Bright, Harold: *MICOM Operational Baseline Cost Estimate Concept Implementation*, MICOM, Huntsville, AL, 1985.

Bullard, John W.: *History of Redstone Missile System*, Army Missile Command, Huntsville, AL, October 1965, AMC 23 M.

Burkey, Martin: "NASA Faulted for Space Station Automation Plans," *Huntsville Times*, Huntsville Times, Huntsville, AL, November 16, 1986.

Burt, David N.: "Proactive Procurement: The Key to Increased Profits," *Productivity and Quality*, Prentice-Hall, Englewood Cliffs, NJ, 1984.

Center for Strategic and International Studies: *U.S. Defense Acquisition: A Process in Trouble*, Georgetown University, Washington, DC, March 1987.

Cerf, Christopher, and Henry Beard: *The Pentagon Catalog*, Workman Publishing Company, New York, 1986.

Cibinic, John, Jr., and Ralph C. Nash, Jr.: *Administration of Government Contracts*, George Washington University, Washington, DC, 1981.

Cibinic, John, Jr., and Ralph C. Nash, Jr.: *Cost Reimbursement Contracting*, George Washington University, Washington, DC, 1981.

Cibinic, John, Jr., and Ralph C. Nash, Jr.: *Formation of Government Contracts*, George Washington University, Washington, DC, 1982.

Clark, John Russell: *Federal Spending Practices, Efficiency, and Open Government*, John Russell Clark, Dallas, TX, May 1973.

Cleland, David I., and Harold Kerzner: *The Best Managed Projects*, Procurement Associates, Covina, CA, April 1986, GCS 8-86 D-1.

Cohen, William A.: *How to Sell to the Government*, Wiley, New York, 1981.

Commerce Clearing House: *Government Contracts Reports: Armed Services Procurement Regulation Manual*, Commerce Clearing House, Chicago, IL, September 1975, ASPM No. 1.

Commission on Government Procurement: *Listing of Study Topics from Approved Study Group Work Plans*, Commission on Government Procurement, Washington, DC, August 1971.

Commission on Government Procurement: *Major Systems Acquisition—Final Report*, Commission on Government Procurement, Washington, DC, January 1972.

Commission on Government Procurement: *Special Report on Acquisition of Major Systems—Draft*, Commission on Government Procurement, Washington, DC, December 1972.

Commission on Government Procurement: *Summary of the Report of the Commission on Government Procurement*, Commission on Government Procurement, Washington, DC, December 1972.

Comptroller General of US: *Managing the Cost of Government*, General Accounting Office, Washington, DC, February 1985, AFMD-85-35.

Conahan, Frank C.: *The B-1B Aircraft Program* (testimony to Armed Services Commission), General Accounting Office, Washington, DC, February 1987, T-NSIAD-87-4A.

Corps of Engineers: *Program Projections*, Headquarters, Corps of Engineers, Washington, DC, October 1986.

Costello, Robert B.: *DOD Response to GAO/NSIAD-87-140*, Department of Defense, Washington, DC, September 1987, P/CPF.

DARCOM: *This is DARCOM*, DARCOM, Alexandria, VA, September 1983, P-360-1.

DARCOM: *Operational Baseline Cost Estimate (OBCE)*, DARCOM, Alexandria, VA, June 1984.

DARCOM: *Development of Maturing US Army Operational Baseline Cost Est (RFP)*, DARCOM, Alexandria, VA, July 1986.

Davidson, Roger H., and Thomas Kephart: *Indicators of Senate Activity and Workload*, Congressional Research Service, Washington, DC, June 1985, 85-133 S.

Davidson, Roger H., and Thomas Kephart: *Indicators of House of Representatives Workload and Activity*, Congressional Research Service, Washington, DC, June 1985, 85-136 S.

Davidson, Roger H., and Walter J. Oleszek: *Congress and Its Members*, CQ Press, Washington, DC, 1985.

Defense Logistics Studies (DLSIE): *Custom Bibliography Comparative Systems*, Defense Logistics Studies, Fort Lee, VA, January 1985.

Defense Systems Management College: *Program Manager*, Defense Systems Management College, Fort Belvoir, VA, July-August 1983, Vol. XII, No. 4.

Defense Systems Management College: *Program Manager*, Defense Systems Management College, Fort Belvoir, VA, January-February 1984, Vol. XIII, No. 1.

Department of the Air Force: *Air Force Acquisition Statement FY87*, Department of the Air Force, Arlington, VA, 1986.

Department of Defense: *Report of the Acquisition Cycle Task Force—Executive Summary*, Department of Defense, Washington, DC, March 1978.

Department of Defense: *Guide to Resources and Sources of Information for Acquisition Research*, Department of Defense, Washington, DC, January 1980.

Department of Defense: *Preparation and Review of Selected Acquisition Reports*, Department of Defense, Washington, DC, May 1980, 7000.3-G.

Department of Defense: *Revision of DOD Instruction 5000.2 Major System Acquisition Procedure*, Department of Defense, Washington, DC, March 1983.

Department of Defense: *Guidance on the Acquisition Improvement Program (AIP)—Memorandum*, Department of Defense, Washington, DC, May 1983.

Department of Defense: *Defense Logistics Studies Information Exchange*, Department of Defense, Fort Lee, VA, September 1983.

Department of Defense: *Defense Financial and Investment Review*, Department of Defense, Washington, DC, 1985.

Department of Defense: *Defense Management Education and Training Catalog*, Department of Defense, Washington, DC, July 1985, DOD 5010.16C.

Department of Defense: *Program Acquisition Costs by Weapon System*, National Technical Information Service, Springfield, VA, February 1986, PB86-153004.

Department of Defense: *Baselining of Selected Major Systems—Directive*, Department of Defense, Washington, DC, August 1986, 5000.45.

Department of Defense: *Management Improvement Plan FY1987-1988*, Department of Defense, Washington, DC, August 1986.

Department of Energy: *Financial Reporting System Validation (RFP)*, Department of Energy, Washington, DC, 13 August 1980.

Department of the Navy: *Best Practices How to Avoid Surprises in World's Most Complicated Procedures*, Department of the Navy, March 1986, NAVSO P-6071.

Doke, Marshall J., Jr. (Editor): *Developments in Government Contract Law—1978*, American Bar Association, Chicago, IL, 1982.

Dominguez, Raymond: *The Selected Acquisition Report (SAR) Rev 2*, Assistant Secretary of Defense, Washington, DC, November 1979.

Driessnack, Lt. Gen. Hans H.: "Defense—But Not at Any Cost," *Program Manager*,

Defense Systems Management College, Fort Belvoir, VA, July-August 1983, Vol. XII, No. 4.

English, Robert J.: *Federal Government Subcontract Forms*, Callaghan, Wilmette, IL, 1983.

Farnsworth, George L.: "HUD Implements Commercial Accounting System," *Government Computer News*, Government Computer News, Silver Springs, MD, November 1986.

Federal Acquisition Institute: *Principles of Government Contract Law*, Federal Acquisition Institute, Washington, DC, September 1979.

Federal Contract Management: A Manual for the Contract Professional, M. Bender, New York, 1982.

Ferris, Nancy: *Inman Sees Turf Battles Ruling Procurement Policy, Government Computer News*, Government Computer News, Silver Springs, MD, November 21, 1986.

Fettig, Lester: "Defense Management Reform: The Carlucci Initiatives a Year Later," *Military Electronic/Countermeasures*, PASHA Publications, Arlington, VA, May 1982.

Financial Accounting Foundation: *Annual Report Financial Accounting Standards Board*, Financial Accounting Foundation, Stamford, CT, 1985.

Flippo, Ronnie: *Federal Financial Management Improvement Act*, Office of Ronnie Flippo, Washington, DC, August 1986.

Flippo, Ronnie: *Procurement Legislation in Process*, Office of Ronnie Flippo, Washington, DC, August 1986.

Fox, John Ronald: *Arming America: How the US Buys Weapons*, Harvard University, Boston, MA, 1974.

Friedman, Milton, and Rose Friedman: *Free to Choose*, Avon Books, New York, 1980.

Gansler, Jacques S.: *The Defense Industry*, MIT Press, Cambridge, MA, 1980.

General Accounting Office: *Theory and Practice of Cost Estimating for Major Acquisitions*, General Accounting Office, Washington, DC, July 1972, B-163058.

General Accounting Office: *Cost Growth in Major Weapon Systems*, General Accounting Office, Washington, DC, March 1973, B-163058.

General Accounting Office: *Recommendations on Executive Branch Progress and Status: Commission on Government Procurement*, General Accounting Office, Washington, DC, January 1974, B-160725.

General Accounting Office: *Executive Branch Actions on Recommendations of Commission on Government Procurement*, General Accounting Office, Washington, DC, 1976, PSAD-76-39.

General Accounting Office: *Legislative Recommendations of Commission on Government Procurement 5 Years Later*, General Accounting Office, Washington, DC, July 1978, PSAD-78-100.

General Accounting Office: *Impediments to Reducing the Costs of Weapon Systems*, General Accounting Office, Washington, DC, November 1979, PSAD-80-6.

General Accounting Office: *Effectiveness of US Forces Can be Increased Through Improved Weapon Design*, General Accounting Office, Washington, DC, January 1981, PSAD-81-17.

General Accounting Office: *Acquiring Weapon Systems in Period of Rising Expenditures*, General Accounting Office, Washington, DC, May 14, 1981, MASAD-81-26.

General Accounting Office: *Improving the Weapon Systems Acquisition Process—Letter*, General Accounting Office, Washington, DC, May 15, 1981, MASAD-81-29.

General Accounting Office: *Consolidated Space Operations Center Lacks Adequate DOD Planning*, General Accounting Office, Washington, DC, January 1982, MASAD-82-14.

General Accounting Office: *Army's Multiple Launch Rocket System is Progressing Well and Merits Support*, General Accounting Office, Washington, DC, February 1982, MASAD-82-13.

General Accounting Office: *Status of Major Acquisitions 30 Sep 81 Better Reporting Essential*, General Accounting Office, Washington, DC, April 1982, MASAD-82-24.

General Accounting Office: *Improving Effectiveness and Acquisition Management of Selected Weapons Systems*, General Accounting Office, Washington, DC, May 1982, MASAD-82-34.

General Accounting Office: *Assessment of Admiral Rickover's Recommendations*, General Accounting Office, Washington, DC, 1983, Vol. 263, A57.

General Accounting Office: *Metro Needs to Better Manage Its Railcar Procurement*, General Accounting Office, Washington, DC, August 1983, NSIAD-83-26.

General Accounting Office: *Status of Major Acquisitions as of Sep 30, 1982*, General Accounting Office, Washington, DC, September 1983, NSIAD-83-32.

General Accounting Office: *Joint Major System Acquisition by the Military Services: An Elusive Strategy*, General Accounting Office, Washington, DC, December 1983, NSIAD-84-22.

General Accounting Office: *DOD Needs to Provide More Credible Weapon Systems Cost Estimates to Congress*, General Accounting Office, Washington, DC, May 1984, NSIAD-84-70.

General Accounting Office: *Managing the Cost of Government Building Financial Management Structure*, General Accounting Office, Washington, DC, February 1985, AFMD-85-35.

General Accounting Office: *Recommendations on Department of Defense Operations*, General Accounting Office, Washington, DC, February 1985, OIRM-85-2.

General Accounting Office: *Hazards of Beginning Weapons Systems Production without Testing*, General Accounting Office, Washington, DC, June 1985, NSIAD-85-68.

General Accounting Office: *Why Some Weapon Systems Encounter Production Problems*, General Accounting Office, Washington, DC, 1985, 85-603178.

General Accounting Office: *Weapons Acquisition Processes of Selected Foreign Governments*, General Accounting Office, Washington, DC, February 1986, NSIAD-86-51FS.

General Accounting Office: *Improving Operations of Federal Departments and Agencies*, General Accounting Office, Washington, DC, February 1986, OIRM-86-1.

General Accounting Office: *Strengthening the Capabilities of Key Acquisition Personnel*, General Accounting Office, Washington, DC, May 1986, NSIAD-86-45.

General Accounting Office: *Weapon Systems Issues Concerning Army's Ligh Helicopter Program*, General Accounting Office, Washington, DC, May 1986, NSIAD-86-121.

General Accounting Office: *Software Project Army Materual Command Spent Millions Without Knowing Total Costs and Benefits*, General Accounting Office, Washington, DC, June 1986, IMTEC-86-18.

General Accounting Office: *US Antisatellite Program Responses to Questions on Program Funding*, General Accounting Office, Washington, DC, June 1986, NSIAD-86-161BR.

General Accounting Office: *SDI Status of Airborne Optical Adjunct and Terminal Imaging Radar*, General Accounting Office, Washington, DC, June 1986, NSIAD-86-153.

General Accounting Office: *Test Resources Early Testing of Major ASW Weapons Can be Enhanced*, General Accounting Office, Washington, DC, June 1986, NSIAD-86-174.

General Accounting Office: *Budget Issues: Cost Escalation on 3 Major DOT Projects*, General Accounting Office, Washington, DC, July 1986, AFMD-86-31.

General Accounting Office: *Air Traffic Controller FAA's Advanced Automation System Acquisition Strategy is Risky*, General Accounting Office, Washington, DC, July 1986, IMTEC-86-24.

General Accounting Office: *Air Traffic Control Status of FAA's Host Computer Project and Software Enhancements*, General Accounting Office, July 1986, IMTEC-86-25BR.

General Accounting Office: *DOD Acquisition Case Study of Army Tactical Missile System*, General Accounting Office, Washington, DC, July 1986, NSIAD-86-45S2.

General Accounting Office: *DOD Acquisition Case Study of Navy V-22 Osprey Joint Vertical Lift*, General Accounting Office, Washington, DC, July 1986, NSIAD-86-45S-7.

General Accounting Office: *DOD Acquisition Case Study Air Force Advanced ASM Program*, General Accounting Office, Washington, DC, July 1986, NSIAD-86-45S-11.

General Accounting Office: *DOD Acquisition Case Study of Air Force Advanced Warning System*, General Accounting Office, Washington, DC, July 1986, NSIAD-86-45S-14.

General Accounting Office: *DOD Acquisition Case Study of MILSTAR Satellite Communications System*, General Accounting Office, Washington, DC, July 1986, NSIAD-86-45S-15.

General Accounting Office: *DOD Acquisition Case Study Air Force Small Intercontinental Ballistic Missile*, General Accounting Office, Washington, DC, July 1986, NSIAD-86-45S-16.

General Accounting Office: *DOD Acquisition Case Study Air Force Space Based Space Surveillance System*, General Accounting Office, Washington, DC, July 1986, NSIAD-86-45S-17.

General Accounting Office: *Procurement Select Acquisition Report: Suggested Approaches for Improvement*, General Accounting Office, Washington, DC, July 1986, NSIAD-86-118.

General Accounting Office: *Missile Develop AMRAAM Certification Issues*, General Accounting Office, Washington, DC, July 1986, NSIAD-86-124BR.

General Accounting Office: *Evaluating the Acquisition and Operation of Information Systems*, General Accounting Office, Washington, DC, July 1986, Tech Guide 2.

General Accounting Office: *Listing of Recent Defense Acquisition Studies—Bibliography*, General Accounting Office, Washington, DC, August 22, 1986.

General Accounting Office: *Selected Civilian Agencies' Cost Estimating Processes for Large Projects*, General Accounting Office, Washington, DC, September 1986, GGD-86-137BR.

General Accounting Office: *Status of the Defense Acquisition Improvement Program's 33 Initiatives*, General Accounting Office, Washington, DC, September 1986, NSIAD-86-178BR.

General Accounting Office: *Advantages and Disadvantages of Centralized Civilian Acquisition Agency*, General Accounting Office, Washington, DC, November 1986, NSIAD-87-36.

General Accounting Office: *Weapon Performance: Operational Test and Evaluation Can Contribute More to Decisionmaking*, General Accounting Office, Washington, DC, December 1986, NSIAD-87-57.

General Accounting Office: *Contract Pricing/Defense Contractor Cost Estimating Systems*, General Accounting Office, Washington, DC, June 1987, NSIAD-87-140.

General Accounting Office and Auditor of Canada: *Federal Government Reporting Study—Summary Report*, General Accounting Office, Washington, DC, 1984, AFMD-86-30.

General Accounting Office and Auditor of Canada: *Illustrated Annual Financial Report: U.S. Government*, General Accounting Office, Washington, DC, 1984, AFMD-86-30A.

General Services Administration: *United States Government Manual 1987/1988*, General Services Administration, Washington, DC, 1987.

Goodwin, Jacob: *Brotherhood of Arms*, Times Books, New York, 1985.

Gossom, W.J.: *Control of Projects, Purchasing, and Materials*, Penn Well Books, Tulsa, OK, 1983.

Gottlieb, Daniel W.: "Former Air Force Purchasing Chief Calls for Civilian Help (Harry Page)," *Purchasing*, Cahners Publishing Company, Boston, March 1986.

Government Counselling Ltd.: *Taurus Online Puts Government Procurement Regulations at Fingertips*, Government Counselling Ltd., Springfield, VA, 1985.

Greenly, Robert B.: *How to Win Government Contracts*, Van Nostrand-Reinhold, New York, 1983.

Gregory, William H.: "Long Awaited Response on Procurement," *Aviation Week*, McGraw-Hill, New York, July 1985, Vol. 123, No. 15.

Hardeman, William E.: "Saving 10 Billion per Year on Parts Competitively," *American Metal Market*, Fairchild Publishing Inc., New York, March 1985, Vol. 93, No. 18.

Harr, Karl G., Jr.: *Letter to Richard D. DeLauer on Procurement Reforms*, Aerospace Industries Association of America, Washington, DC, November 5, 1981.

Hart, Gary, and William S. Lind: *America Can Win*, Adler & Adler, Bethesda, MD, 1986.

Headquarters, Air Force Systems Command: *The Affordable Acquisition Approach Study—Executive Summary*, Headquarters, Air Force Systems Command, Andrews AFB, MD, February 1983.

Hess, Stephen: *Organizing the Presidency*, Brookings Institution, Washington, DC, 1976.

Hiestand, O. S., Jr.: "A Doubter's Assessment of the Packard Commission Report," *Contract Management*, National Contract Management Association, McLean, VA, September 1986.

Hirsch, William J.: *The Contracts Management Deskbook*, American Management Association, New York, 1983.

House Armed Services Committee: *Aspin, Mavroules Have Proposal to End Inflation "Lottery"—News Release*, House Armed Services Committee, Washington, DC, August 3, 1986.

House Armed Services Committee: *Cost of Salt Breakout Could Hit $100 Billion over Decade—News Release*, House Armed Services Committee, Washington, DC, August 4, 1986.

House Armed Services Committee: *ASPIN Says GOP Also Repudiating Reagan on Arms Control—New Release*, House Armed Services Committee, Washington, DC, August 15, 1986.

House Armed Services Committee: *Dickinson Says Floor Votes Reek of Hypocrisy—News Release*, House Armed Services Committee, Washington, DC, August 19, 1986.

House Armed Services Committee: *Aspin Says Air Force Seems to be Editing the Law—News Release*, House Armed Services Committee, Washington, DC, August 21, 1986.

Huzel, Dieter K.: *Peenemunde to Canaveral*, Prentice-Hall, Englewood Cliffs, NJ, 1962.

Iacocca, Lee, and William Novak: *Iacocca: An Autobiography*, Bantam Books, New York, 1984.

Industrial Capacity and Defense Planning, Lexington Books, Lexington, MA, 1983.

Institute of Cost Analysis: *Proceedings of the First Annual ICA Symposium*, Institute of Cost Analysis, Washington, DC, December, 1982.

Institute of Cost Analysis: *Journal of Cost Analysis*, Institute of Cost Analysis, Washington, DC, Spring 1985, Vol. 2.

Institute of Cost Analysis: "Newsletter of the Institute," *Newsletter of the Institute*, Institute of Cost Analysis, Alexandria, VA, August 1986, Vol. II, No. 2.

International Systems: *PAC III: The Complete Project Management System*, International Systems, King of Prussia, PA.

Ireland, Lewis R.: *Project Management: Critical Success Factors and Keys to Effectiveness*, SWL, Inc., McLean, VA, August 1986, 58-07-86.

Jane's Publishing Co.: *Jane's Catalog and Defense Weekly*, Jane's Publishing Co., New York, 1986.

Joint Financial Management Improvement Program: *Proceedings of the Fifteenth Annual Financial Management Conference 1986*, Joint Financial Management Improvement Program, Washington, DC, March 1986.

Jordan, John: *Illustrated Directory of Modern American Weapons*, Prentice-Hall, Englewood Cliffs, NJ, 1986, 86-12295.

Kaplan, Fred: *Pentagon is Missing Up to $32 Billion, Study Says, Huntsville Times*, Huntsville Times, Huntsville, AL, September 3, 1986.

Kennedy, John J.: *Incentive Contracts and Cost Growth*, Air Force Business Research Management Center, Wright-Patterson AFB, OH, October 1983.

Kennedy, John J.: *The Effectiveness of the Award Fee Concept*, Naval Office for Acquisition Research, Washington, DC, June 1986.

Keyes, W. Noel: *Keyes' Encyclopedic Dictionary of Procurement Law, 2nd ed.*, Oceana Publications, Dobbs Ferry, NY, 1982.

Lambro, Donald: "Pentagon Mis-Management Costs Taxpayers Billions," *Dollars and Sense*, National Taxpayers Union, Washington, DC, September 1986, Vol. 17, No. 7.

Large Scale Programs Institute (LSPI): *Colloquium on Research Priorities for Large Scale Programs 1985-1*, Large Scale Programs Institute, Austin, TX, March 1985.

Laughlin, Edward P.: *Education Plan for NES*, National Estimating Society, Washington, DC, January 1986.

Lauria, Lew S.: *NASA Financial Management Systems*, NASA, Washington, DC, December 1986.

Livingston, John C., and Robert G. Thompson: *The Consent of the Governed*, Macmillan, New York, 1963.

Management Sciences Institute: *Financial Management Courses*, Management Sciences Institute, Philadelphia, PA, 1982, Bull. No. 78.

Marshall Space Flight Center, NASA: *Hubble Space Telescope*, Marshall Space Flight Center, Huntsville, AL, 2M684.

McNichols, Gerald R. (Editor): *Cost Analysis*, Operation Research Society of America, Arlington, VA, 1984.

McVay, Barry L.: *Getting Started in Federal Contracting*, Panoptic Enterprises, Woodbridge, VA, 1984.

Military Business Publishers: *Military Business Review*, Military Business Publishers, Virginia Beach, VA, July-August 1986, Vol. 2, No. 4.

Miller, Richard K. (Editor): *Handbook on Selling to the U.S. Military*, Fairmont Press, Atlanta, GA, 1983.

Mitchell, J. E.: *Communication Segment Overview*, Spacecom, Gaithersburg, MD, January 1985.

Mobile Data Services: *Systems Analysis Rept for Communications and Supt Segments of CSOC*, CACI, Gaithersburg, MD, February 1985.

Naisbitt, John: *Megatrends*, Warner Books, New York, 1982.

Naisbitt, John, and Patricia Aburdene: *Re-inventing the Corporation*, Warner Books, New York, 1985.

NASA: *Guidelines for Evaluation of Contractor Accounting Systems*, NASA, Washington, DC, February 1967, NHB 9090.6.

NASA: *Management Study of NASA Acquisition Process Report of Steering Group*, NASA, Washington, DC, 1971.

NASA: *Procedures for Contractor Reporting of Correlated Cost and Performance Data*, NASA, Washington, DC, October 1971.

NASA: *Source Evaluation Board Manual*, NASA, Washington, DC, December 1975, NHB 5103.6A.

NASA: *Principles of Project Management*, NASA, Washington, DC, January 1982, NHB 7120.2.

NASA, Marshall Space Flight Center: *Specifications and Standards Approved Baseline List*, Marshall Space Flight Center, NASA, Huntsville, AL, October 1982, MM 8070.2D.

NASA: *Marshall Space Flight Center 1960–1985 Anniversary Report*, U.S. Government Printing Office, Washington, DC, 1985, 25M1285.

NASA: *NASA Facts*, U.S. Government Printing Office, Washington, DC, September 1985, NF-143.

Nathan, Richard P.: *The Plot that Failed: Nixon and the Administrative Presidency*, Wiley, New York, 1975.

National Commission on Space: *Pioneering the Space Frontier*, Bantam Books, New York, May 1986.

National Contract Management Association: *Contract Management*, National Contract Management Association, McLean, VA, July 1986, Vol. 26, Issue 7.

National Contract Management Association: *Contract Management*, National Contract Management Association, McLean, VA, October 1986, Vol. 26, Issue 10.

National Contract Management Association: *Contract Management*, National Contract Management Association, McLean, VA, December 1986, Vol. 26, Issue 12.

National Contract Management Association: *Contract Management*, National Contract Management Association, McLean, VA, February 1987, Vol. 27, Issue 2.

National Defense Headquarters: *DND Economic Model*, National Defense Headquarters, Ottawa, Ontario, May 1982.

National Estimating Society: *National Estimating Society Conference '85*, National Estimating Society, Washington, DC, June 1985.

National Estimating Society: *National Estimator*, National Estimating Society, Washington, DC, Fall 1985, Vol. 6, No. 3.

National Estimating Society: "Cost Plus Contracts," *Dollars and Sense*, National Estimating Society, Washington, DC, June 1986.

National Estimating Society: "Pentagon Inspector General's Report Rebuts Navy on Aircraft Costs," *Dollars and Sense*, National Estimating Society, Washington, DC, August 1986.

National Journal, Inc.: *Government Executive*, National Journal, Inc., Washington, DC, March 1987, Vol. 19, No. 3.

National Military Publications: *Huntsville Salutes Redstone Arsenal*. National Military Publications, El Cajon, CA, 1986.

91st Congress: *Commission on Government Procurement Public Law 91-129 83 Stat 269*, 91st Congress, Washington, DC, November 1969, PL 91-129.

98th Congress, 2d Session, House of Representatives: *Defense Spare Parts Procurement Reform Act*, House of Representatives, Washington, DC, April 1984, 98-690.

Nozette, Stewart, and Barney Roberts: *LSPI/NASA Workshop in Lunar Base Methodology Development—Final Report*, Large Scale Programs Institute, Austin, TX, November 1985, NAG9-116.

Office of Federal Procurement Policy: *Major System Acquisitions Discussion of OMB*, Office of Federal Procurement Policy, Washington, DC, August 1976, OFPP No. 1, Circ. No. A-109.

Office of Federal Procurement Policy: *Survey and Study of Executive Agency Procurement Regulations*, Office of Federal Procurement Policy, Washington, DC, April 1979.

Office of Federal Procurement Policy: *Proposal for Uniform Federal Procurement System*, Office of Federal Procurement Policy, Washington, DC, February 1982.

Office of Federal Procurement Policy: *The Federal Acquisition Regulation (FAR) Planning Document*, Office of Federal Procurement Policy, Washington, DC, December 1982, Vol. 1, pp. 1–51.

Office of Federal Procurement Policy: *Activities of the Office of Federal Procurement Policy*, Office of Federal Procurement Policy, Washington, DC, January-December 1983.

Office of Federal Procurement Policy: *Activities of the Office of Federal Procurement Policy*, Office of Federal Procurement Policy, Washington, DC, January-December 1984.

Office of Federal Procurement Policy: *Competition in the Award of Subcontracts by Fed Prime Contractors FY82*, Office of Federal Procurement Policy, Washington, DC, April 1984.

Office of Federal Procurement Policy: *Review of the Procurement Actions of DOD*. Office of Federal Procurement Policy, Washington, DC, April 1984.

Office of Federal Procurement Policy: *Review of the Spare Parts Procurement Practices of DOD*, Office of Federal Procurement Policy, Washington, DC, June 1984.

Office of Federal Procurement Policy: *Competition in Contracting Act of 1984 Analysis*, Office of Federal Procurement Policy, Washington, DC, October 1984.

Office of Federal Procurement Policy: *Procurement Policy Letters*, Office of Federal Procurement Policy, Washington, DC, September 1985, OFPP No. 6.

Office of Management and Budget: *Major System Acquisitions*, Office of Management and Budget, Washington, DC, April 1976, Circ. No. A-109.

Office of Management and Budget: *Report to Congress 1975 Office of Federal Procurement Policy*, Office of Management and Budget, Washington, DC, 1976.

Office of Management and Budget: *Report to the Congress 1976 Office of Federal Procurement Policy*, Office of Management and Budget, Washington, DC, May 1977.

Office of Management and Budget: *Executive Order 12352 on Federal Procurement Reforms: A Reform '88 Program*, Office of Management and Budget, Washington, DC, February 1984, M-84-7.

Office of Management and Budget: *Budget of the United States Government FY87, 4 vols.*, Office of Management and Budget, Washington, DC, 1986.

Office of Management and Budget: *Joint Financial Management Improvement Program (JFMIP): Annual Report 85*, Office of Management and Budget, Washington, DC, 1986.

Office of Management and Budget: *United States Government Standard General Ledger*, Office of Management and Budget, Washington, DC, August 1986.

Ordway, Frederick I., and Mitchel Sharpe: *The Rocket Team*, MIT Press, Cambridge, MA, 1982.

Packard Commission: *A Formula for Action*, Packard Commission, Washington, DC, April 1986.

Packard Commission: *A Quest for Excellence Final Report of Packard Commission*, 2 vols., Packard Commission, Washington, DC, June 1986.

Packard Commission: *Conduct and Accountability*, Packard Commission, Washington, DC, June 1986.

Packard Commission: "U.S. Defense Procurement," *Journal of Defense Diplomacy*, Packard Commission, McLean, VA, July 1986.

Pascale, Richard T., and Anthony G. Athos: *The Art of Japanese Management*, Warner Books, New York, 1981.

Peck, Merton, J., and Frederic M. Scherer: *The Weapons Acquisition Process: An Economic Analysis*, Harvard University, Boston, MA, 1962.

Peters, Thomas J., and Robert H. Waterman, Jr.: *In Search of Excellence*, Harper & Row, New York, 1982.

Presidential Commission on Space Shuttle: *Report of the Presidential Commission on Space Shuttle Challenger Accident*, Presidential Commission on Space Shuttle, Washington, DC, June 1986.

Preston, Colleen A.: "Congress and the Acquisition Process: Some Recommendations for Improvement," *National Contract Management Journal*, National Contract Management Association, McLean, VA, Summer 1986, Vol. 20, Issue 1.

Price, Douglas, S.: "Golden: Let's Beef Up Financial Systems," *Government Computer News*, Government Computer News, Silver Spring, MD, July 4, 1986.

Price Waterhouse: *Enhancing Government Accountability*, Price Waterhouse, New York, 1983.

Project Management Institute: *Project Management Journal*, Project Management Institute, Drexel Hill, PA, August 1984, 0147-5363.

Proxmire, William: "Why Military Contracting is Corrupt," *New York Times*, New York Times, New York, December 1985, Vol. 135, No. 15.

Reagan, Ronald: *Executive Order 12352 on Federal Procurement Reforms*, The White House, Washington, DC, March 17, 1982.

Revel, Jean-François: *How Democracies Perish*, Harper & Row, New York, 1983.

Rich, Michael, Edmund Dews with C. L. Batten, Jr.: *Improving the Military Acquisition Process*, Rand Corporation, Santa Monica, CA, February 1986, R-3373-AF/RC.

Riley, Dennis J.: *Federal Contracts, Grants and Assistance*, Shepard's/McGraw-Hill, Colorado Springs, CO, 1983.

Rishe, Melvin: *Government Contract Costs*, Federal Publications, Washington, DC, 1984.

Ritchey, L. S.: *ASD/Industry Cost Estimating Workshop 22-24 Oct 85*, U.S. Air Force, Dayton, OH, December 1985.

Ropelewski, Ross R.: "Navy Restructures Acquisition Process," *Aviation Week*, McGraw-Hill, New York, April 1985, Vol. 122, No. 15.

Rose, Richard: *Managing Presidential Objectives*, Macmillan, New York, 1976.

Schuller, Robert H., and Paul David Dunn: *The Power of Being Debt Free*, Thomas Nelson, Nashville, TN, 1985.

Shea, David: *Information from Department of Air Force*, Department of Air Force, Andrews AFB, MD, September 1986.

Shelkin, Michael, and J. Philip Purdy: "The Counterobstacle Vehicle," *The Military Engineer*, Society of American Military Engineers, Alexandria, VA, July 1986, No. 508.

Sherman, Stanely N.: *Government Procurement Management, 2nd ed.*, Wordcrafters Publications, Gaithersburg, MD, 1985.

Sichnitzer, Paul A.: *Government Contract Bidding, 2nd ed.*, Federal Publications, Washington, DC, 1982.

Sink, D. Scott, Sandra J. DeVries, and Thomas C. Tattle: "An In-Depth Study and Review of State-of-the-Art and Practice Productivity Measurements" *IE Management News*, AIIE, Norcross, GA, Winter 1985, Vol. XIX, No. 2.

Society of American Value Enginners: *Value World*, Society of American Value Engineers, Chicago, IL, October-December 1985, Vol. 8, No. 3.

Spigarelli, Lt. Col. Raymond F.: "Multinational Source Selection," *Program Manager*, Department of Defense, Ft Belvoir, VA, May-June 1982, Vol. XI, No. 3.

Srinivasan, Raghavan, and David M. Sassoon: *International Contracting and Procurement for Development Projects*, International Law Institute, Washington, DC, 1982.

Stewart, Rodney D.: *Major Systems Planning and Budgeting in Support of National Goals*, George C. Marshall Space Flight Center, NASA, Huntsville, AL, 1972.

Stewart, Rodney D.: *Comments on OMB Circular A-109 Major Systems Acquisition*, Commission on Government Procurement, Washington, DC, January 1973.

Stewart, Rodney D., and Ann L. Stewart: *Proposal Preparation*, Wiley, New York, 1984.

Stewart, Rodney D., and Richard M. Wyskida (Editors): *Cost Estimator's Reference Manual*, Wiley, New York, 1987.

Stilkind, Jerry: "MLRS Stands Out as a Procurement Success," *Army Times*, Army Times, Arlington, VA, November 25, 1985.

"Streamlining Planning and Programming Phase," *Aviation Week*, March 1985, Vol. 122, No. 25.

Technology Reviews: "Interview: Richard DeLauer on Defense," *Technology Review*, July 1986, Vol. 89, No. 5.

Television Program: It's Your Business: *Highways USA: What's the Right Route?*, U.S. Chamber of Commerce, Washington, DC, February 21, 1987, Program 390.

Textron: *Textron Annual Report 1985*, Textron, Providence, RI, 1985.

Thamhain, Hans J.: *Engineering Program Management*, Wiley, New York, 1984.

Time, Inc.: "The Navy Under Attack," *Time*, Time, Inc., New York, May 8, 1978, Vol. III, No. 19.

Time, Inc.: "Fat on the Sacred Cow," *Time*, Time, Inc., New York, February 22, 1982.

Time, Inc.: "Shots Heard Across the Atlantic," *Time*, Time, Inc., New York, March 1, 1982.

Time, Inc.: "The Winds of Reform," *Time*, Time, Inc., New York, March 7, 1983.

Tobias, Sheila, Peter Goudinoff, Stefan Leader, and Shelah Leader: *What Kinds of Guns are They Buying for your Butter?*, William Morrow, New York, 1982.

Trueger, Paul M.: *Accounting Guide for Defense Contracts, 7th ed.*, Commerce Clearing House, Chicago, IL, 1982.

Uken, Duane H.: *BMDATC Technology Base MIS—Final Report*, Engineering & Economics Research, Huntsville, AL, June 15, 1984.

U.S. Air Force Aeronautical System Division: *Cost Estimating Handbook—Simulators*, U.S. Air Force Aeronautical System Division, Wright Patterson AFB, OH, August 1981, 82-506-HU.

U.S. Army Materiel Command: *AMC—Providing Leaders the Decisive Edge*, U.S. Army Materiel Command, Huntsville, AL, February 1986.

U.S. Army Missile Command: *History of the U.S. Army Missile Command 1962–1977*, U.S. Army Missile Command, Huntsville, AL, 1979, DARCOM 84M.

U.S. Army Missile Command: *Multiple Launch Rocket System*, U.S. Army Missile Command, Huntsville, AL, April 1984.

U.S. Congress: *Hearings on Federal Spending Practices, Efficiency, and Open Government*, U.S. Congress, Washington, DC, 1975.

U.S. Congress, House: *1984 Competition in Contracting Act*, U.S. Congress, House, Washington, DC, 1984, 98-1157.

U.S. Congress House: *Omnibus Procurement Reform Amendment to HR 4428, DOD Authorization*, U.S. Congress (House), Washington, DC, August 1986.

U.S. Congress, House, Armed Services: *National Defense Authorization Act for Fiscal Year 1987*, U.S. Congress, House, Washington, DC, July 1986, 99-718.

U.S. Congress, Senate: *Major Systems Acquisition Reform Hearings*, U.S. Congress, Senate, Washington, DC, 1975, 61-499.

U.S. Congress, Senate: *Reducing Cost of Weapon System Acquisition*, U.S. Congress, Senate, Washington, DC, 1985, 85-602128.

Watson, Gregory H.: *Marketing Research and Development Concepts to the Navy*, Contintential Publishing, McLean, VA, 1985.

Webster, Francis M., Jr.: *Survey of Project Management Software Packages*, Project Management Institute, Drexel Hill, PA, October 1985.

"White House Calls for Changes in Procurement Procedures," *Aviation Week*, McGraw-Hill, New York, November 1985, Vol. 123, No. 25.

Index